Alexander Schilling

Model-Based Detection and Isolation of Faults of Diesel Engines

Alexander Schilling

Model-Based Detection and Isolation of Faults of Diesel Engines

The Lambda and the Nitrogen Oxides Sensors as Instruments to Detect and Isolate Faults in the Air and Fuel Paths of Common-rail DI Diesel Engines

Südwestdeutscher Verlag für Hochschulschriften

Impressum/Imprint (nur für Deutschland/ only for Germany)

Bibliografische Information der Deutschen Nationalbibliothek: Die Deutsche Nationalbibliothek verzeichnet diese Publikation in der Deutschen Nationalbibliografie; detaillierte bibliografische Daten sind im Internet über http://dnb.d-nb.de abrufbar.

Alle in diesem Buch genannten Marken und Produktnamen unterliegen warenzeichen-, marken- oder patentrechtlichem Schutz bzw. sind Warenzeichen oder eingetragene Warenzeichen der jeweiligen Inhaber. Die Wiedergabe von Marken, Produktnamen, Gebrauchsnamen, Handelsnamen, Warenbezeichnungen u.s.w. in diesem Werk berechtigt auch ohne besondere Kennzeichnung nicht zu der Annahme, dass solche Namen im Sinne der Warenzeichen- und Markenschutzgesetzgebung als frei zu betrachten wären und daher von jedermann benutzt werden dürften.

Verlag: Südwestdeutscher Verlag für Hochschulschriften Aktiengesellschaft & Co. KG
Dudweiler Landstr. 99, 66123 Saarbrücken, Deutschland
Telefon +49 681 37 20 271-1, Telefax +49 681 37 20 271-0, Email: info@svh-verlag.de
Zugl.: Zürich, ETH, Diss., 2008

Herstellung in Deutschland:
Schaltungsdienst Lange o.H.G., Berlin
Books on Demand GmbH, Norderstedt
Reha GmbH, Saarbrücken
Amazon Distribution GmbH, Leipzig
ISBN: 978-3-8381-0687-8

Imprint (only for USA, GB)

Bibliographic information published by the Deutsche Nationalbibliothek: The Deutsche Nationalbibliothek lists this publication in the Deutsche Nationalbibliografie; detailed bibliographic data are available in the Internet at http://dnb.d-nb.de.

Any brand names and product names mentioned in this book are subject to trademark, brand or patent protection and are trademarks or registered trademarks of their respective holders. The use of brand names, product names, common names, trade names, product descriptions etc. even without a particular marking in this works is in no way to be construed to mean that such names may be regarded as unrestricted in respect of trademark and brand protection legislation and could thus be used by anyone.

Publisher:
Südwestdeutscher Verlag für Hochschulschriften Aktiengesellschaft & Co. KG
Dudweiler Landstr. 99, 66123 Saarbrücken, Germany
Phone +49 681 37 20 271-1, Fax +49 681 37 20 271-0, Email: info@svh-verlag.de

Copyright © 2009 by the author and Südwestdeutscher Verlag für Hochschulschriften Aktiengesellschaft & Co. KG and licensors
All rights reserved. Saarbrücken 2009

Printed in the U.S.A.
Printed in the U.K. by (see last page)
ISBN: 978-3-8381-0687-8

Preface

This thesis is based on my research on diesel engines performed at the Measurement and Control Laboratory (IMRT) of the ETH Zurich between 2003 and 2008.

First of all, I would like to express my gratitude to my advisor, Prof. Dr. Lino Guzzella, for proposing this project and for providing me outstanding support. I really appreciated the close collaboration, and the positive working environment resulting from his trust.

Furthermore, I would like to extend my thanks to Prof. Dr. Isermann for accepting to be my first co-examiner, and for his professional contributions.

A special thank you goes to my second co-examiner, Dr. Alois Amstutz, for his valuable support, for the fruitful discussions, and for being a constant source of constructive inputs to the work.

Many thanks also to Dr. Chris Onder for his availability for discussions, and for his incessant help in finding always the best solutions.

Then, I would like to gratefully acknowledge the financial support by the Forschungsvereinigung Verbrennungskraftmaschinen e.V. (FVV) in Frankfurt a. Main, Germany, and the Federal Office for the Environment (FOEN) in Switzerland. A special thank is made to the research group "Emission-Controlled Diesel Engines", which was initiated by Dr. Rainer Buck of Robert Bosch GmbH, first leaded by Dr. Klaus Allmendinger, and successively by Dipl.-Ing. Zandra Jansson of Daimler AG in Stuttgart, Germany; it supported the project in terms of content and brought in the interest of the industrial partners.

I owe a lot to the staff of the IMRT, in particular to: Hansueli Honegger, Oskar Brachs, and Jan Prikryl for providing decisive technical support for the test-bench setup; Brigitte Rohrbach for carefully reviewing this text and earlier publications, and for always giving useful comments that improved my written English; Claudia Wittwer and Annina Müller for their help in managing the administrative issues.

Very special thanks go to all the past and present colleagues at IMRT, whit whom I spent a lot of interesting and enjoyable hours. Among them: Michael Benz, Yves Hohl, Ezio Alfieri, Charlie Boston, Raphaël Suard, Dr. Daniel Brand, Dr. Mikael Bianchi, Dr. Antonio Sciarretta, Dr. Marco Gerig, Matt Donovan, and Søren Ebbesen.

My gratitude goes also to the staff of the Laboratory for Aerothermochemistry and Combustion System (LAV) of ETH Zurich, in particular to Fabrizio Noembrini and Patrick Kirchen, for the many fruitful discussions and for supplying the cylinder pressure analysis tool WEG.

Finally, I would like to express my overwhelming gratitude to my parents, Maria and Peter Schilling, as well as to my sister Isabella Schilling, for allowing me to study at the ETH, and consequently to complete this work. The last very, very special thanks go to my girlfriend Luana, whose support and encouragement made the realization of this dissertation possible.

Alexander Schilling
Zurich, June 2008

Contents

Abstract	vii
Zusammenfassung	ix
List of Symbols	xi

1 Introduction — 1
   1.1 Background and Motivation 1
   1.2 Objectives, Proposed Approach, and Contributions . . 8
   1.3 Outline 10

2 Experimental Facilities — 13
   2.1 Test-Bench Setup 13
   2.2 Development Environment 14
   2.3 Solid-State Emission Sensors 15
      2.3.1 Wide-Band λ Sensor 17
      2.3.2 Nitrogen Oxides Sensor 18

3 Combustion Model — 21
   3.1 Motivation and Previous Work 21
   3.2 Model Description 23
      3.2.1 Injection 24
      3.2.2 Heat Release 26
      3.2.3 Single-Zone Engine Process 29
      3.2.4 Reaction Temperature 34
      3.2.5 Emissions 36

	3.3	Model Calibration	40
	3.4	Model Verification	45
4	**Control-Oriented Emission Models**		**51**
	4.1	Motivation and Previous Work	51
	4.2	Virtual Sensors for λ and NOx	54
		4.2.1 Description of the Models	54
		4.2.2 Calculation of the Model Parameters	59
		4.2.3 Example	62
	4.3	Adaptive Virtual NOx Sensor	67
		4.3.1 Adaptation as a Prerequisite to FDI	68
		4.3.2 Choice of the Parameters to be Adapted	70
		4.3.3 Adaptation Algorithm	70
		4.3.4 Example	75
	4.4	Application to a Series-Production Engine	79
5	**Model-Based Fault Detection and Isolation**		**85**
	5.1	Contribution	85
		5.1.1 Previous Work	86
		5.1.2 Proposed Work	89
	5.2	Fault Detection	90
	5.3	FDI Strategy A	91
		5.3.1 Fault Estimation and Classification	93
	5.4	FDI Strategy B	98
		5.4.1 Fault Estimation	98
		5.4.2 Fault Classification	101
6	**Test on the NEDC**		**107**
	6.1	Measurements	107
		6.1.1 The Driving Cycle	107
		6.1.2 Inverse λ and PM Emissions	108
		6.1.3 Fault Emulation	108
	6.2	Results	112
		6.2.1 Control-Oriented Emission Models	112
		6.2.2 FDI Strategy A	113

	6.2.3 FDI Strategy B	113
6.3	Discussion	125

7 Conclusions — 129

A Combustion Model: Equations and Parameters — 131
- A.1 Injection — 131
- A.2 Heat Release — 132
  - A.2.1 Evaporation of the Injected Fuel — 132
  - A.2.2 Air/fuel Mixture Preparation — 135
  - A.2.3 Ignition Delay — 137
  - A.2.4 Premixed Combustion — 138
  - A.2.5 Diffusion Combustion — 140
  - A.2.6 Superposition of Premixed and Diffusion Combustion — 142
- A.3 Single-Zone Engine Process — 142
  - A.3.1 Cylinder Process — 142
  - A.3.2 Cylinder Wall Heat Loss — 144
  - A.3.3 Air and Exhaust Gas Properties — 147
  - A.3.4 Exhaust Gas Recirculation — 149
  - A.3.5 Gas Exchange — 150
- A.4 Reaction Temperature — 154
- A.5 Emissions — 155
  - A.5.1 Nitrogen Oxides — 155
  - A.5.2 Soot — 160

B Extended Kalman Filter — 163
- B.1 Description and Assumptions — 163
- B.2 Algorithm — 164
- B.3 Tuning — 166
- B.4 Observability and Excitation — 166

C Instrumentation — 169
- C.1 Exhaust Gas Analyzers — 169
- C.2 Sensors — 171

Bibliography 173

Abstract

The present work deals with the diagnostics problem of diesel engines. Until now, to this author's knowledge, no diagnostics system for modern diesel engines based on emission sensors has been proposed. This work can thus be considered as a first attempt to make good use of a wide-range sensor for the relative air/fuel ratio (Lambda, λ) and of a nitrogen oxides concentration sensor to realize a diagnostics system able to detect, estimate, and classify faults of the air and fuel paths of modern diesel engines.

First, a model of the crank-angle discrete combustion process of the test-bench diesel engine is developed. It is to help the understanding of the NO_x emission formation process and to provide the basics for the further development of the control-oriented models. The model is calibrated by means of static measurements and a complete cylinder pressure analysis.

Next, two control-oriented emission models for the real-time prediction of the NO_x emission and the relative air/fuel ratio are developed. These models are based on a linear approach. The parameters needed are calculated efficiently by means of simulations, using the previously developed combustion model as a virtual engine.

According to the physical definition of the relative air/fuel ratio, the linear approach is justified in the case of the corresponding control-oriented model, but not in the case of the control-oriented NO_x model, where that linear approach represents only an approximation of the real behavior, which is known to be non-linear. This problem reduces the effective accuracy of the control-oriented NO_x model. Hence an adaptation algorithm is developed in order to slightly adjust certain parameters of the control-oriented NO_x model. The aims are to increase its accuracy and to match the characteristics of each individual engine.

Finally, both control-oriented models are used together with sensors for the relative air/fuel ratio and NO_x emission for the development of the two proposed fault detection and isolation strategies A and B. These are then able to detect and isolate faults of the injected fuel quantity, the air mass flow, and the boost pressure.

All models and algorithms presented are tested by means of experiments conducted on an engine test-bench, and the results obtained are presented and discussed.

Zusammenfassung

Die vorliegende Arbeit behandelt das Thema der Diagnose von modernen Dieselmotoren. Nach der Kenntnis des Autors liegen bisher keine auf Emissionssensoren basierenden Systeme zur Diagnose von Dieselmotoren vor. Diese Arbeit kann somit als erster Versuch gesehen werden, zur Diagnose eine Breitband-Lambda-Sonde und einen Stickoxydsensor zu verwenden. Das präsentierte Diagnosesystem ist in der Lage, Fehler in den Luft- und Kraftstoffpfaden von modernen Dieselmotoren zu detektieren, zu schätzen und zu klassifizieren.

In einem ersten Schritt wurde direkt aus physikalischen Überlegungen ein detailliertes Verbrennungsmodell hergeleitet. In diesem Modell wurden alle relevanten Prozesse, die während einer üblichen dieselmotorischen Verbrennung vorkommen, modelliert. Die Kalibrierung des detaillierten Verbrennungsmodells erfolgte anhand von statischen Messungen sowie einer vollständigen Analyse des Zylinderdruckverlaufs.

Dann wurden zwei lineare regelungstechnische Emissionsmodelle für die echtzeitfähige Berechnung der NO_x-Emissionen und des Luftverhältnisses entwickelt. Das Verbrennungsmodell diente als virtueller Motor, um die Parameter dieser Modelle schnell und effizient mittels Simulationen zu bestimmen.

Gemäss der physikalischen Bedeutung des Luftverhältnisses ist der verwendete lineare Ansatz im Falle des regelungstechnischen Luftverhältnismodells gerechtfertigt. Im Falle des regelungstechnischen NO_x-Modells stellt der lineare Ansatz hingegen nur eine Approximation dar. Diese Tatsache reduziert die Genauigkeit dieses Modells. Auf diesem Grund wurde ein Adaptionsalgorithmus entwickelt, um einige Parameter des NO_x-Modells leicht anzupassen. Das Ziel dieses Vorgehens ist es, die Genauigkeit des regelungstechnischen NO_x-Modells zu erhöhen, und das Modell in Übereinstimmung mit den Charakteristiken jedes einzelnen Motors zu bringen.

Basierend auf den regelungstechnischen Emissionsmodellen und auf Informationen aus den Lambda- und NO_x-Sensoren wurden zwei Fehlererkennungssysteme A und B für die Detektion, die Schätzung und die Klassifizierung von Fehlern in der Einspritzung, in der Luftmenge und im Ladedruck entwickelt.

Alle Modelle und Algorithmen wurden mittels Experimenten am Prüfstand erprobt und die erhaltenen Resultate werden hier vorgestellt und analysiert.

List of Symbols

Acronyms

Symbol	Description
A/F	air/fuel
AIR	air mass flow
AP	adapted parameters
BMEP	brake mean effective pressure
CA	crank-angle
CLD	chemiluminescence detector
CR	common-rail
DI	direct injection
ECT	erroneous classification time
ECU	electronic control unit
EGR	exhaust gas recirculation
EKF	extended Kalman filter
EVO	exhaust valve opening
FD	fault detection
FDI	fault detection and isolation
FDT	fault detection time
HC	hydrocarbons
HEGO	heated exhaust gas oxygen (sensor)
HFM	hot-film mass flow sensor
HiL	hardware-in-the-loop
ID	ignition delay
INJ	injected fuel quantity
IPSO	intake port shut-off

IVC	intake valve closing	
LNT	lean NOx trap	
nAP	non-adapted parameters	
NEDC	new European driving cycle	
NN	neural network	
NOx	nitrogen oxides	
OBD	on-board diagnosis	
PASS	photo-acustic soot sensor	
PIM	boost pressure	
PM	particulate matter	
QSS	quasi-static simulation	
RLS	recursive least-square	
RT	real-time	
SCR	selective catalytic reduction	
SI	spark-ignited	
SMD	Sauter mean diameter	
SOC	start of combustion	
SOI	start of injection	
VNT	variable nozzle turbine	

Symbols Latin

Symbol	Description	Unit
a	discrete sensor parameter	−
A	discrete dynamics matrix	−
A	surface	m^2
b	discrete sensor parameter	−
B	discrete input matrix	−
B_v	discrete input noise matrix	−
c_d	discharge coefficient	−
c_m	mean piston speed	m/s
C	discrete output matrix	−
d, D	diameter	m
f	discrete linear system function	−

List of Symbols

g	discrete linear output function	—
h	specific enthalpy	J/kg
H_f	fuel lower heating value	J/kg
k	discrete time variable	—
l	length	m
L	Kalman gain	—
m	mass	kg
n	number of ...	—
n	moles	mol
n_d	discrete gas transport and sensor delay	—
n_e	engine speed	rpm
p	pressure	Pa
q	heat flow	W/m^2
q	discrete measurement noise vector	—
Q	heat	J
Q	residual covariance matrix	—
r	radius	m
r	residuals	—
s	speed	m/s
R	gas constant	$J/(kg \cdot K)$
$\tilde{R}$	universal gas constant	$J/(mol \cdot K)$
R_q	measurement noise covariance matrix	—
R_v	input noise covariance matrix	—
t	time	s
T	temperature	K
T_{sample}	sample time	s
u	velocity	m/s
u	specific internal energy	J/kg
u	model input vector	—
U	internal energy	J
v	discrete input noise vector	—
V	volume	m^3
V_{inj}	volume of fuel injected in cylinder	$mm^3/stroke$
w	engine input vector	—

W	work	J
x	state vector	–
y	model and engine output vector	–
z	discrete Laplace variable	–

Symbols Greek

Symbol	Description	Unit
Δ	absolute difference	–
δ	relative difference	–
χ_{st}	stoichiometric A/F ratio	–
η	efficiency	–
φ	discrete crank-angle variable	rad
$\Delta\varphi_{sample}$	sample crank-angle	rad
κ	isentropic exponent	–
λ, Λ	relative A/F ratio	–
$\mu\sigma$	variable discharge coefficient	–
ω	revolution speed	rad/s
Π	pressure ratio	–
ρ	density	kg/m^3
Σ	covariance matrix	–
θ	parameter	–
τ	characteristic time	s
τ_d	gas transport and sensor delay time	s

Superscripts

Symbol	Description
0	reference state
m	measured values
$\wedge$	EKF estimated values
$*$	reference values
–	modified vectors and matrices for the EKF

List of Symbols

Subscripts

Symbol	Description
0	reference
air	air
$aireg$	air + exhaust gas
$avail$	available
b	burned
bb	blow-by
$bore$	bore
c	compression
cyl	cylinder
d	displaced
$diff$	diffusion
dr	droplet
e	engine, equilibrium
eff	effective
ex	exhaust
f	fuel
gc	gas convection
gex	gas exchange
gl	global
ic	intercooler
in	intake, inflowing
inj	injection
lam	laminar
liq	liquid
max	maximum
noz	nozzle
out	outflowing
pr	particle radiation
pre	premixed
r	reaction

$rail$	rail
ref	reference
str	stroke
th	theoretical
tot	total
$turb$	turbulent
u	unburned
v	valve
vap	evaporated
vol	volumetric
w	cylinder wall
$zone$	premixed zone
λ	concerning the λ sensor and model
N	concerning the NO_x sensor and model

Chapter 1

Introduction

The whole work is briefly introduced, pointing out the motivation, the aims, and the strategy pursued.

1.1 Background and Motivation

Since the beginning of the 1990s diesel engines have benefited from a noticeable technological evolution, made possible through the development of new injection systems, such as common-rail (CR), which allow the direct injection (DI) of the fuel in the combustion chamber with very high pressures (up to 2000 bar) as well as the free shaping of the injection profile. Furthermore, the application of an exhaust gas recirculation (EGR) system and a variable nozzle turbocharger (VNT) have succeeded in lowering the emissions levels while still improving the driving dynamics. The control of all these new degrees of freedom is a very challenging task, as described in [4] and [46].

In order to fulfill legislation requirements, namely low levels of nitrogen oxides (NO_x) and particulate matter (PM), as well as to meet customer desires (driving dynamics), highly sophisticated measurement devices and control algorithms have been developed and implemented in the electronic control unit (ECU) of diesel engines. Much work has been done on conventional diesel engine control strategies, see [22], [70], [83], [120], [122], as well as on more advanced ones, see for example [51], [84], [93], [94], [97], [108], [110].

However, in the future the compliance of diesel engines with increasingly stringent emission regulations, Table 1.1, will no longer be possible with the optimization of the combustion process alone, see [56]. Rather, the application of novel control and diagnostics strategies based on the knowledge of the engine-out information will become necessary, together with improved exhaust gas aftertreatment systems.

Table 1.1: Existing and proposed European emission standards for diesel passenger cars in g/km (source: www.dieselnet.com/standards/eu/ld.php).

stage	date	NOx	PM	durability
EURO III	2000	0.50	0.05	80000 km or 5 years
EURO IV	2005	0.25	0.025	100000 km or 5 years
EURO V	2009	0.18	0.005	160000 km or 5 years
EURO VI	2014	0.08	0.005	160000 km or 5 years

This approach leads to at least three conceivable scenarios for the development of future diesel engines, which are described below.

Scenario 1: Feedback Control of Emissions

One possibility is that diesel engines will become feedback controlled systems. Despite the technological level reached by modern DI diesel engines, one fundamental problem is still unresolved: Engines are basically feedforward controlled systems, as shown in the upper graph of Fig. 1.2, i.e., the control of the air and fuel paths is actually performed on the basis of sensors placed before the engine process (air mass flow sensor, boost pressure sensor, and rail pressure sensor). For example, in the case of the EGR control loop, the actuator is represented by the EGR valve and the sensor by the air mass flow sensor. This feedforward control approach implies that the effects on emissions of deteriorated engine parts, e.g. due to ageing, cannot be compensated

1.1. Background and Motivation

by the control system, but rather have to be taken into account preventively when designing the engine. This is an unavoidable precaution, because each engine has to meet the emission standards during a long period of time as shown in Table 1.1. Obviously, the drawback is that engines must be designed with sufficiently large emission tolerances, causing the efficiency to be lower, and thus the consumption to be higher than it could theoretically be. Figure 1.1 illustrates this problem qualitatively. This problem has been known for a long time. First attempts to realize a feedback control for diesel engines were published as early as 1990. Figure 1.2 shows the comparison between a typical arbitrary control loop for diesel engines and a possible emission-based feedback control system. Probably the first contribution is [23], where a control system for a diesel engine with a Comprex charger based on a switching-type relative air/fuel ratio (λ) sensor is proposed. That work was taken up and expanded upon [41], where a control system for a turbocharged diesel engine based on a wide-range λ sensor is described. The authors of [89] present a similar system. In [49], a control strategy for heavy-duty diesel engines based on the estimated values of the NO_x and soot emissions is presented. More recently, a cylinder pressure based control system for diesel engines using individual control of fuel injection timing and quantity was proposed in [92], where also results from investigations on the potential for vehicle emissions dispersion reduction were presented.

Scenario 2: Emission-Based Exhaust Gas Aftertreatment Systems Control

Another probable scenario is the introduction of aftertreatment systems, such as catalytic converters based on the selective catalytic reduction (SCR) process or lean NO_x traps (LNT) for the abatement of the NO_x emissions. In an SCR system, the NO_x gases are removed by means of chemical reactions involving a reductant, normally urea or ammonia, which is added to the exhaust gas and absorbed onto a catalyst. The products of the reaction are water vapor and nitrogen gas. The control of an SCR process is a challenging task, see [35],

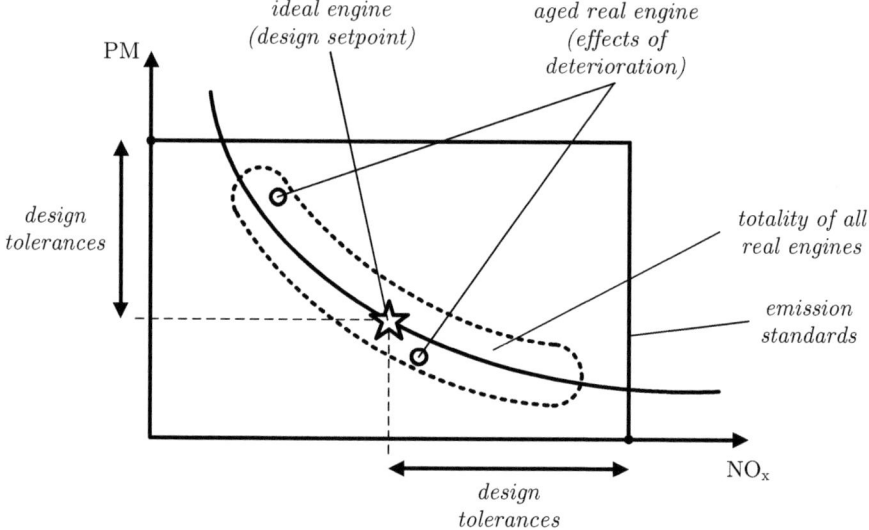

Figure 1.1: Qualitative effects of engine ageing on the NO_x and PM emissions, and design tolerances of current modern diesel engines.

[112], and Fig. 1.3, since the exact quantity of reductant to be injected and the exact moment of injection have to be determined in order to avoid ammonia slip (too much ammonia injected) or NO_x slip (too little ammonia injected). In an LNT system, a NO_x adsorber based on a catalytic converter support, coated with a special washcoat containing zeolites, traps the NO_x molecules acting as a "molecular sponge." Once the adsorber is full it has to be regenerated [77]. Since this process normally takes places every 30 to 90 s, the challenge is to determine as exactly as possible when the absorber is full of NO_x molecules.

Scenario 3: Emission-Based Monitoring and Diagnosis

The increasingly stringent limitations on emission levels imply more narrow tolerances of operations, such that diesel engines have to be continuously monitored in order to ensure the optimality of the ac-

1.1. Background and Motivation

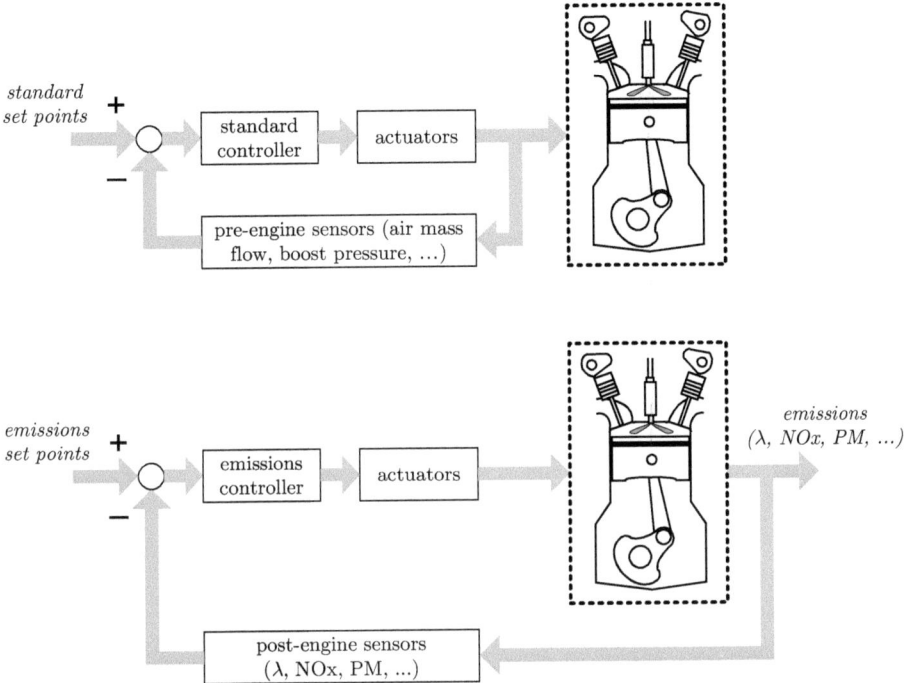

Figure 1.2: Comparison between a standard arbitrary control loop for current diesel engines (top), and a possible emission-based feedback control system (bottom).

tual operating conditions. Moreover, the more stringent on-board diagnosis (OBD) thresholds that will be applied with future emission regulations force the development of more sophisticated diagnosis concepts for OBD systems, which could be emission-based. Figure 1.4, for instance, schematically shows the fault detection and isolation (FDI) concept of the present work. The scope of an OBD system is to identify malfunctions and deteriorations of the engine and aftertreatment system components that cause emissions to exceed certain thresholds fixed by the legislation, see Table 1.2. The driver is informed about failures by a malfunction indicator light (MIL). Furthermore, the ve-

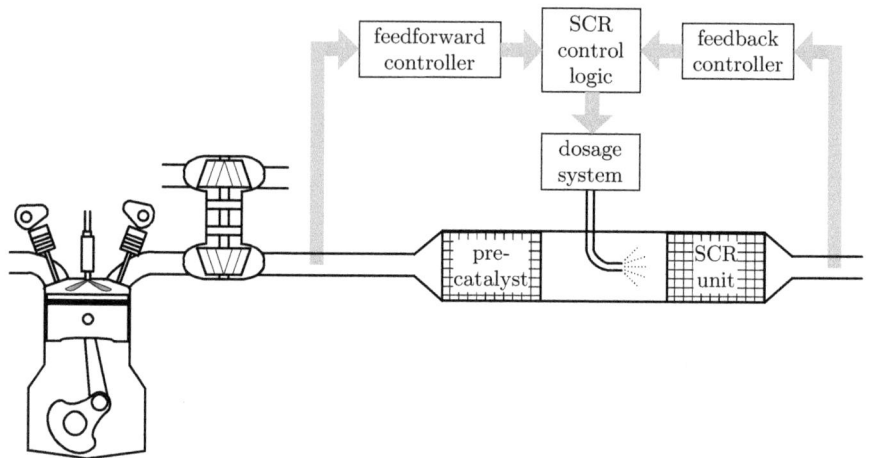

Figure 1.3: The control strategy for an SCR system according to [35] and [112].

hicle is equipped with a standardized fast digital communications port to provide real-time data in addition to a standardized series of diagnostic trouble codes. The current legislation for OBD systems of diesel passenger cars prescribes the monitoring of the catalytic converter, the PM trap, the fuel injection system, the emission control system components, and the other emission-related components. Hence major damage to the engine and aftertreatment systems can be avoided, and the pollutant emissions are always under control. Moreover, OBD systems yield a fundamental support for technical inspections, servicing, periodic monitoring, and repair of the vehicle.

In all of these cases, i.e., feedback emission control in Scenario 1, emission-based aftertreatment control in Scenario 2, and novel emission-based diagnostics algorithms for OBD systems in Scenario 3, the knowledge of the engine outputs is a fundamental prerequisite. This knowledge could be gained either with real sensors or with virtual ones, i.e., with mathematical representations of the various processes describing the emissions formation that can be implemented in the ECU of an en-

1.1. Background and Motivation

Table 1.2: European OBD threshold limits for diesel passenger cars in g/km (source: www.dieselnet.com/standards/eu/ld.php).

stage	date	NOx	PM
EURO III	2003	1.20	0.18
EURO IV	2005	1.20	0.18

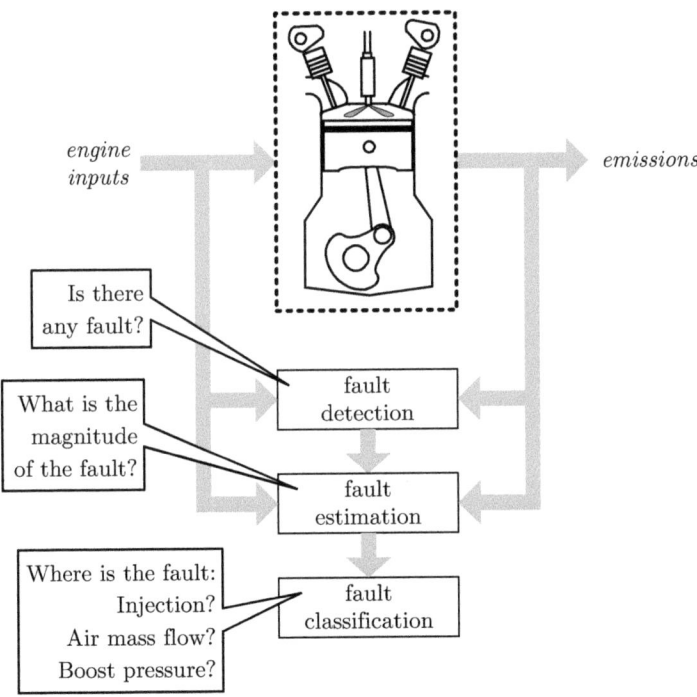

Figure 1.4: The emission-based FDI concept of the present work as a possible approach for future OBD systems.

gine. Currently, the only engine-output sensors commercially available are those for measuring λ and the NO_x concentration level.

1.2 Objectives, Proposed Approach, and Contributions

The aim of this work is to explore the possibilities given by the application of current solid-state λ and NO_x emission sensors for the detection and isolation of faults in the air and fuel paths of modern CR DI diesel engines. To achieve this objective a model-based strategy is pursued.

In a first step, a mathematical model of the engine is developed. The objective is to obtain a physics-based model with good prediction and extrapolation capabilities in order to perform plausible qualitative and quantitative evaluations of the influence of the different engine inputs on emissions, fuel consumption, and engine performance. The computational burden of this model inhibits its implementation in the ECU of the engine, and consequently its application for control and diagnosis purposes.

Therefore, in a second step, control-oriented models for the real-time (RT) computation of the λ value and the NO_x concentration are derived from the detailed combustion model. These models can be implemented in the ECU of the engines, they show good prediction capabilities, and they are easily transferable to other engines. The accuracy of the NO_x model is further improved by means of an adaptive strategy in order to obtain a fast and reliable prediction of the NO_x emission.[1]

Finally, on the basis of the control-oriented models developed, the FDI system is realized. Here, a new approach is treated that includes the implementation of algorithms based on the information obtained from the newly introduced λ and NO_x emission sensors. Faults of the air mass flow, boost pressure, and injected fuel quantity are taken into account to explain discrepancies of the expected λ and/or NO_x values. Fault detection is performed by comparing the sensor signals to

---

[1] The resulting λ and the adaptive NO_x models are used here as reference models for the model-based FDI system, but they could also be used for further applications like feedback emission control, see [71] and [82], and aftertreatment control.

1.2. Objectives, Proposed Approach, and Contributions

the results of the models estimations, and fault location and size are estimated by means of two different strategies based on the extended Kalman filter (EKF) technique. In the first strategy proposed, one single EKF provides a coordinated estimate of the three engine inputs supposed to be faulty. In the second strategy, each engine input supposed to be faulty is estimated independently by means of a bank of three EKFs, one for each fault mode considered, and the residuals of the different EKFs are representative of the quality of the fault estimations, with the smallest residuals indicating the best one and thus the location of the fault.

The control-oriented models and the FDI system are tested on the basis of real transient measurement from the New European Driving Cycle (NEDC). Results show that the control-oriented models are accurate, and thus appropriate for diagnostics purposes. The FDI system can correctly detect, estimate, and classify faults of the injected fuel quantity, of the air mass flow, and of the boost pressure. The beneficial effects of the adaptation procedure of the NO_x model on the FDI system are also demonstrated.

Figure 1.5 gives an overview about the methodology proposed. The contributions of this work can be summarized as follows:

- a methodology for developing RT λ and NO_x models that uses a detailed physical combustion model to compute the necessary parameters and that is easily transferable to other engine systems;

- a novel algorithm based on the EKF technique that is supported by the solid-state NO_x sensor and that adapts on-line the maps of parameters of the RT NO_x model in order to match each individual production engine and thus improve the accuracy;

- RT models for λ and NO_x, which perform a fast and accurate prediction and which can be used also for other purposes than diagnosis, e.g. aftertreatment control or closed-loop emission control;

- two novel attempts to realize a detection and isolation system of faults in the air and fuel paths based on λ and NO_x sensors.

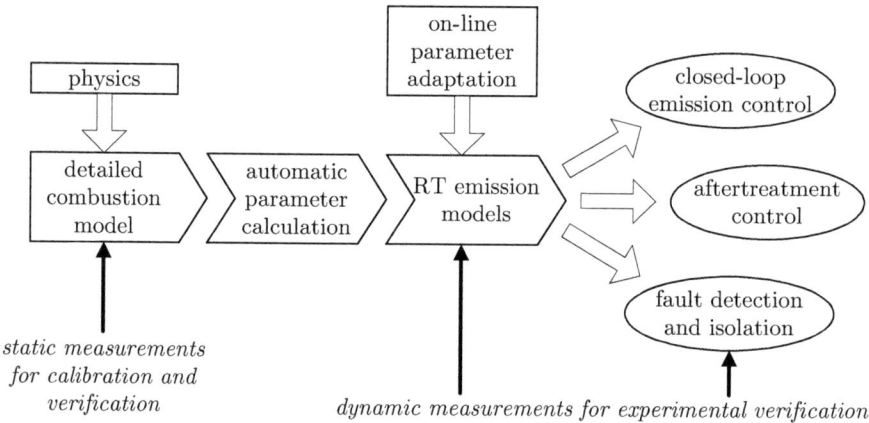

Figure 1.5: The proposed methodology to obtain RT emission models. In this work the models are used for diagnostics purposes, but they could also be used for closed-loop emission control and aftertreatment control.

1.3 Outline

Chapter 2 deals with the "boundary conditions" of the research, i.e, with the experimental facilities used to conduct this work. The test-bench engine is described, as well as all necessary measurement devices. The characteristics of both solid-state λ and NO_x sensors applied for the on-line measurement of emission are also described here.

Chapter 3 describes the development of a detailed physics-based combustion and emission model of a modern diesel engine, including its calibration and verification processes.

Chapter 4 describes the development of the control-oriented emission models used for the FDI system. In particular, Section 4.2 presents

1.3. Outline

the real-time computation of the λ values and the NO_x emission levels obtained by these models and Section 4.3 elicit the adaptive strategy used to improve the accuracy of the NO_x model. In Section 4.4, the whole methodology to derive the adaptive control-oriented NO_x model is re-applied to data obtained from measurements taken on a series-production engine. Results from real measurement data to demonstrate that this methodology is easily transferable and efficacious are also shown.

Chapter 5 describes the development of the FDI system. The fault detection is explained in Section 5.2, the fault estimation and classification strategy based on one single EKF in Section 5.3, and the fault estimation and classification strategy based on the bank of three independent EKFs in 5.4.

Chapter 6 shows results of tests from real NEDC measurements. Results from both control-oriented λ and NO_x models are shown, as well as results from the proposed FDI strategies. Moreover, the positive effects of the adaptation of the NO_x model on the FDI system are discussed here.

Chapter 2
Experimental Facilities

The experimental conditions are described under which the research was performed: The engine, the development environment, the measurement devices, as well as the most important sensors installed on the engine.

2.1 Test-Bench Setup

The measurements and experiments for model calibration and verification purposes for this research are carried out on the modern mass-production diesel engine described in [54] and [65], whose technical data are listed in Table 2.1. A photo of the engine on the test-bench is shown in Fig. 2.1. In the standard configuration, the engine is equipped with an air mass flow sensor and a boost pressure sensor which allow the EGR valve and the position of the VNT vanes to be controlled. Various pressure and temperature sensors are installed for calibration and verification purposes. Additionally, a solid-state wide-range λ sensor and a NO_x sensor are mounted for the on-line measurement of emissions. The complete engine system is schematically shown in Fig. 2.2.

The test-bench is equipped with a dynamic eddy-current brake and with special devices for the measurement of the following emission substances: NO, NO_x, CO, CO_2, HC, and PM. Further technical details about the exhaust gas analyzers are given in Appendix C.

Table 2.1: Technical data of the test-bench engine.

manufacturer	DaimlerChrysler
name	OM 611
type	direct-injection, common-rail
features	EGR, VNT and IPSO
architecture	4 cylinders, 16 valves
bore x stroke	88 mm x 88.4 mm
compression ratio	19:1
maximum torque	300 Nm at 1800-2600 rpm
maximum power	92 kW at 4200 rpm
emissions level	EURO III

2.2 Development Environment

Various software and hardware modules are installed for data acquisition, data processing, engine monitoring, and the implementation of the real-time algorithms. The details are listed in Tables 2.2 and 2.3, respectively. The connection between the various modules is shown in Fig. 2.3.

Table 2.2: The software modules of the development environment.

software module	function
INCA	engine monitoring
ASCET	real-time implementation
dSpace	data acquisition
MATLAB/Simulink	data processing

2.3. Solid-State Emission Sensors

Figure 2.1: The fully instrumented engine on the test-bench.

Table 2.3: The hardware modules of the development environment.

hardware module	function
ES1000	transputer and communication boards
ETK7	interface between ES1000 and ECU
ECU	electronic control unit of the engine
AD/DA	analogical $\leftrightarrow$ digital converter
CAN-bus	signal transmission

2.3 Solid-State Emission Sensors

The main air pollutants emitted by diesel engines are the NO_x and PM. As mentioned in Section 1.1, solid-state NO_x emission sensors are

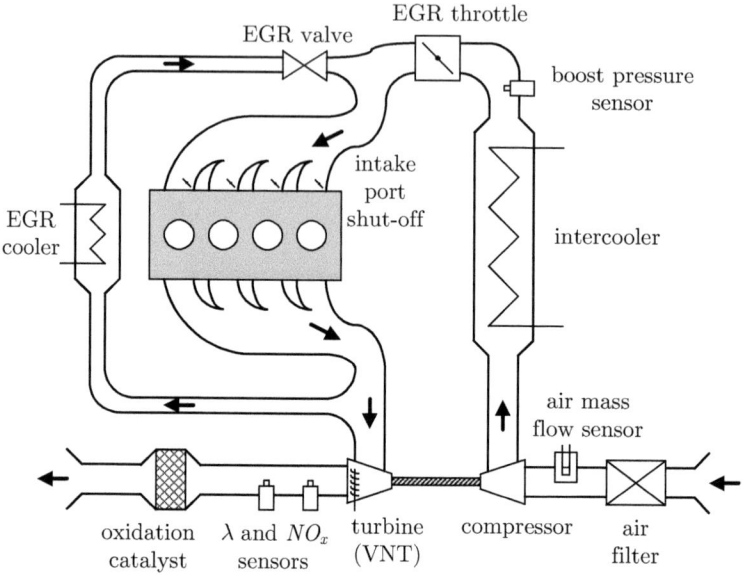

Figure 2.2: Schematic of the complete engine system.

currently commercially available that can be mounted in the tailpipe of diesel engines, and that can be used for real-time purposes. However, there are no such sensors available as yet for the PM. Since the PM emissions are strictly related to the amount of oxygen present during the combustion process, the λ value can be assumed to serve as an indicator for the PM emissions, although it cannot be considered as an "emission" quantity in the strict sense of the term. Hence a wide-band λ sensor is used here as a substitute for the PM sensor. The relationship between λ value and PM emissions is clarified with the results shown in Fig. 2.4, where a simple linear function based on the inverse λ value is tested on the basis of real NEDC measurements in order to describe the instantaneous PM mass emitted, recorded with the PASS-device described in Appendix C.1.

The solutions proposed in this work are based on algorithms supported by information from a λ and a NO_x sensor, thus their main

2.3. Solid-State Emission Sensors

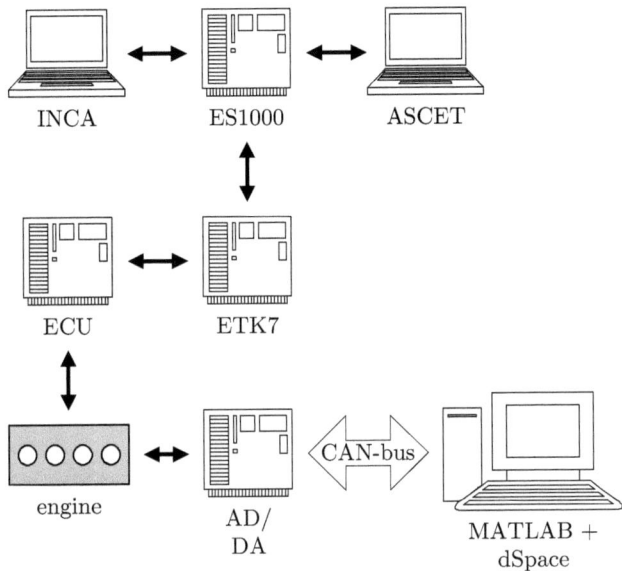

Figure 2.3: The complete development environment: Connections between the various software and hardware modules.

characteristics and properties are explained in the following. Further information about oxygen sensors for automotive applications can be found in [66], [68], and [78].

2.3.1 Wide-Band λ Sensor

The sensor used is a planar wide-band λ sensor with pumped O_2 reference. Its technical data are listed in Table 2.4, and further information can be found in [95]. The λ sensor can be re-calibrated during the fuel cut-off phases where the oxygen concentration is known since only ambient air flows through the engine. This solution is already applied in current engines. The deterioration of the λ sensor is thus not considered here.

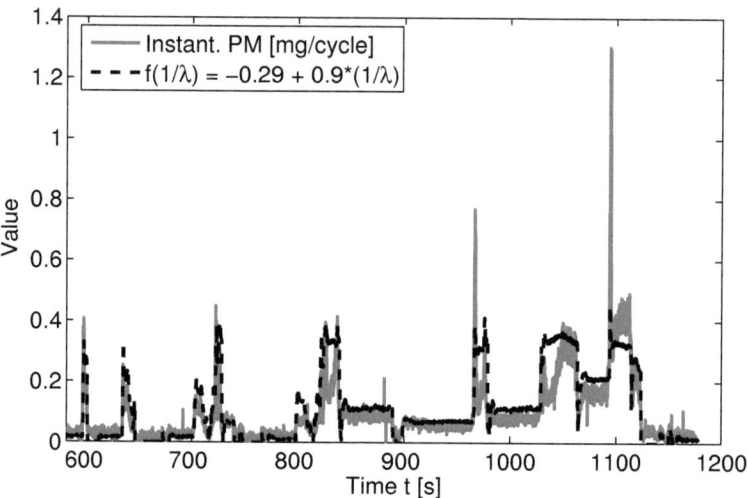

Figure 2.4: Comparison between instantaneous PM mass, measured with the PASS, and a simple linear function based on the inverse λ value for the last half of the NEDC.

Table 2.4: Technical data of the λ sensor used.

range	0.65 - ∞ (air)
accuracy	$\lambda = \begin{cases} 0.8 \Rightarrow \pm 1.3\% \\ 1.0 \Rightarrow \pm 0.6\% \\ 1.7 \Rightarrow \pm 2.9\% \end{cases}$
response time	70 ms

2.3.2 Nitrogen Oxides Sensor

The NO_x sensor used here is the one described in [99], which is composed of a ZrO_2-based multilayer sensor element. The technical data are listed in Table 2.5. Basically, its principle of operation is similar to that of a wide-band λ sensor. Therefore the deterioration of the NO_x sensor is not considered here either.

2.3. Solid-State Emission Sensors

For static operating conditions the performance of this sensor is comparable to that of more sophisticated measurement devices. The left-hand graph in Fig. 2.5 shows the comparison of the measured NO_x levels recorded with the solid-state sensor and a fast CLD NO_x analyzer according to [85] and [90] for different static operating conditions between 1000 and 3500 rpm and between 0 and 16 bar BMEP. The probe of the fast CLD analyzer is mounted in the tailpipe of the test-bench engine near the solid-state sensor.

For transient operating modes the solid-state sensor is too slow to follow the engine dynamics. The following experiment is performed to illustrate the response characteristics of the NO_x sensor. The engine is running at 2000 rpm and 6 bar BMEP, and the start of injection (SOI) signal is anticipated stepwise. Since the SOI has an immediate effect on the NO_x emission, the response of the fast CLD analyzer is also immediate. Thus the signal of the fast CLD NO_x analyzer can be interpreted to represent the engine-out NO_x emission. However, as the graph on the right in Fig. 2.5 shows, in this case the response of the NO_x sensor is affected by a delay (gas transport + sensor) of about 0.6 s and a time constant of about 1 s.

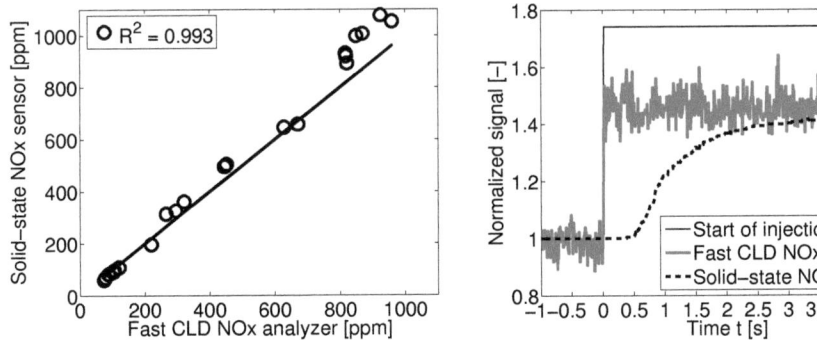

Figure 2.5: Comparison between NO_x concentration levels measured with a fast CLD analyzer and using a solid-state sensor for static operating points (left), and responses of the fast CLD analyzer and of the solid-state sensor after a step of the start of injection signal (right).

This result is in accordance with the specifications of the sensor of Table 2.5 and with the data from [99].

Table 2.5: Technical data of the NO_x sensor used.

range	0 - 1550 ppm
accuracy	$\pm 10\%$
response time	750 ms

Chapter 3

Physical Modeling of the In-Cylinder Combustion Process

A detailed physical model of the combustion process is realized in order to help the understanding of the NO_x emission formation process in diesel engines, and to provide the basics for the further development of the control-oriented models.

3.1 Motivation and Previous Work

To capture all relevant dynamics of the combustion process, a physics-based model is developed. This approach is necessary because the NO_x emissions are strongly dependent on the cylinder pressure and the reaction temperature profiles during the combustion. The model is conceived according to the following guidelines:

- Comprehensiveness ⇒ the effect of all engine variables, describing both the air and the fuel paths, should be captured.

- Simplicity ⇒ a zero-dimensional approach is sufficient, i.e., pressure, temperature, and concentration of substances are assumed to be uniform within the whole control volume.

- Details as needed ⇒ at least a separation into two zones, burned and unburned, is necessary for the chemical description of the NO_x formation.

Since the model is physics-based, simulation results are expected to be accurate, and its application on different engines is to be possible with little effort by virtue of just a few parameters requiring adjustment. Furthermore, the model can be used as a virtual engine to investigate the effect of the engine variables on the emissions. On the other hand, the computational burden involved is such that the model cannot be applied in the ECU of the engines and thus cannot be used for control purposes.

The literature about combustion modeling is exhaustive. Numerous modeling approaches are proposed for various goals and for engine processes with several stages of complexity. The basics are explained in [6], where a comprehensive introduction is given on combustion engines. Moreover, the detailed functionality of diesel engine systems is treated in [2]. Other modeling guidelines for combustion engines are presented in [15], [16], [18], and [19].

Zero-dimensional models of diesel engines are proposed for example in [21], where the model is used as tool for the EGR rate estimation, and in both [22] and [83], where the model is used to develop a model-based control strategy of the VNT and the EGR. In [25] and [86] both a zero-dimensional and a phenomenological model to study the effects of different engine settings on emissions and noise are described. Another phenomenological model is proposed in [31], where the aim is the NO_x and soot computation. In [38] a detailed model implemented and optimized for on-line applications is presented.

3.2 Model Description

Basically, the model can be divided into five sub-models according to the physical causality of the conventional diesel combustion, as Fig. 3.1 clearly shows: Fuel injection, heat release, single-zone cylinder process (inclusive gas exchange), reaction temperature estimation, and emissions formation. These modeling steps are fairly standard, and therefore only a brief description of each corresponding sub-model is given below. Interested readers are referred to the indicated references and to Appendix A, where the complete equation and the parameter set of the complete model implemented in this work are listed.

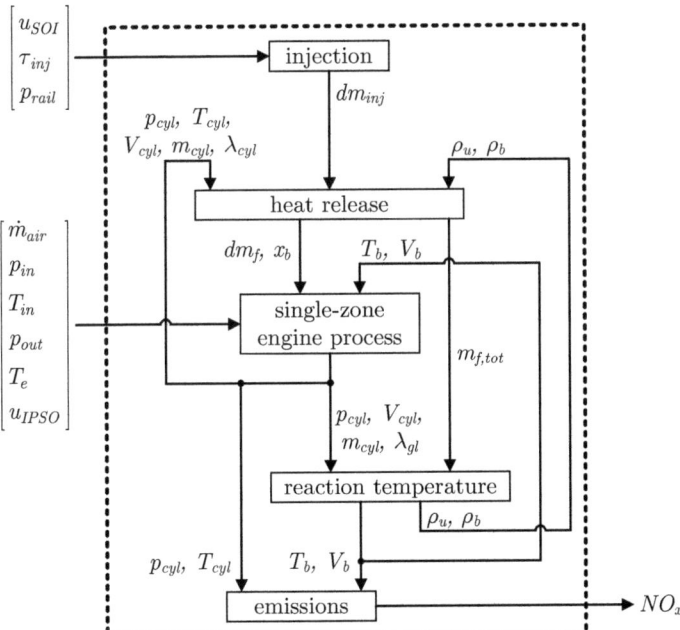

Figure 3.1: The physical causality of the combustion model developed.

The model is developed on a crank-angle discrete basis, because it is a more convenient approach to discretize the engine processes.

Therefore all time derivatives of the various physical equations have to be transformed according to the following relationship:

$$\frac{d\varphi}{dt} = \omega_e \Rightarrow dt = \frac{d\varphi}{\omega_e}. \quad (3.1)$$

The chosen sampling time is $\Delta\varphi_{sample} = 0.4\ CA^o$, which is a compromise between accuracy and calculation time.

3.2.1 Injection

The computational sequence begins with the "injection" sub-model, where the control signals "start of injection" u_{SOI}, "injection duration" τ_{inj}, and "rail pressure" p_{rail} of the injection system are processed in order to obtain the injection rate profile. Various detailed models have been proposed ([29], [87], [107]), as well as models for hardware-in-the-loop (HiL) purposes [119], but the one presented in [25] is chosen for its modeling simplicity and accuracy.

For the complete set of equations and parameters of this sub-model, refer to Appendix A.1.

The basic idea is to describe the injection rate profile as a trapezoid as shown in Fig. 3.2, which is a good approximation of the real profile. Accordingly, two variables have to be calculated to unequivocally define the injection profile for each operating point: The maximum injection rate and the injection rate slope. To do this, reference values for the rail pressure p^0_{rail} and the injection rate slope α^0 are chosen. The corresponding reference maximum injection rate is found using the Bernoulli equation and neglecting the cylinder pressure effect:

$$\frac{dm^0_{inj,max}}{d\varphi} \approx \frac{1}{\omega_e} \cdot c_d \cdot A_{noz} \cdot \sqrt{2 \cdot \rho_{f,inj} \cdot p^0_{rail}}, \quad (3.2)$$

and the maximum injection rate in another operating point than the reference is calculated with the following empirical relation:

$$\frac{dm_{inj,max}}{d\varphi} = \frac{dm^0_{inj,max}}{d\varphi} \cdot \left(\frac{p_{rail}}{p^0_{rail}}\right)^{n_1}, \quad (3.3)$$

3.2. Model Description

where the parameter n_1, which should be 0.5 because of the square root of the Bernoulli equation, is slightly tuned to optimize the model.

The injection rate slope is:

$$\alpha = \frac{\frac{dm_{inj}}{d\varphi}}{\Delta\varphi}, \qquad (3.4)$$

and it is proportional to the rail pressure, according to Newton's Second Law:

$$\underbrace{p_{rail} \cdot A_{noz}}_{\sum F} \propto \underbrace{\alpha \cdot \omega_e^2}_{m \cdot a}. \qquad (3.5)$$

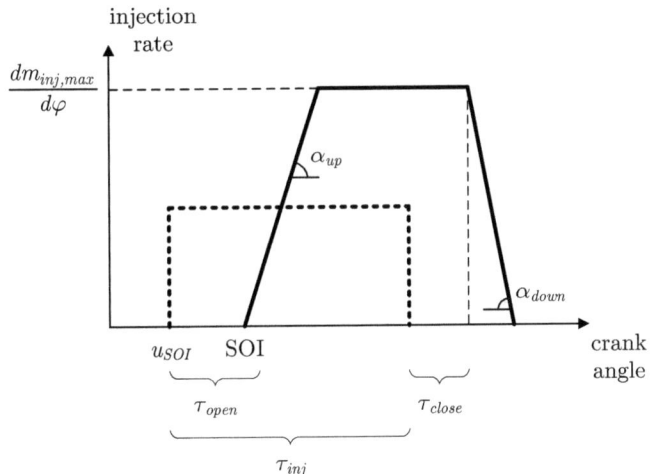

Figure 3.2: Approximation of the real injection rate profile.

Hence the injection rate slope in another operating point than the reference is calculated with the following empirical relation:

$$\alpha = \alpha^0 \cdot \left(\frac{p_{rail}}{p_{rail}^0}\right)^{n_2}, \qquad (3.6)$$

where also the parameter n_2, which should be 1.0 because of the direct proportionality with the rail pressure induced by Newton's Second Law, is slightly tuned to optimize the model.

The parameters τ_{open} and τ_{close} represent the injector opening and closing delay, i.e., the delay time between the control signal of the ECU and the effective movement of the injector. Normally, the injector closing is slightly faster than the injector opening. This fact is taken into account by means of an empirical factor according to the following relation:

$$\alpha_{up} = \alpha \qquad (3.7)$$
$$\alpha_{down} = c \cdot \alpha_{up}, \ c > 1. \qquad (3.8)$$

The model presented is applied to every injection event of the engine. Thus the pilot injection and the main injection are parameterized independently, and the corresponding values for the parameters are different.

3.2.2 Heat Release

The resulting injection rate profile feeds the second sub-model "heat release" where rate of heat release, total converted fuel mass, and combustion advancement are computed. It is a phenomenological model and is completely based on [25]. The model is composed of five parts: Fuel evaporation, air entrainment and fuel-mixture generation, ignition delay, premixed combustion, and diffusion combustion. The pilot injection is assumed to burn exclusively within the premixed mode, whereas the main injection burns within both the premixed and the diffusion modes.

For the complete set of equations and parameters of this sub-model, refer to Appendix A.2.

At the beginning only one single jet of injected liquid fuel is considered (one single jet corresponds to the flow through one single nozzle hole). The corresponding injection rate profile is discretized, and at each time step the amount of evaporated fuel is computed according to the so called "d^2 law," which describes the temporal decrease of the liquid fuel droplets diameter due to evaporation:

$$d_{dr}^2 = d_{dr,liq}^2 - \beta \cdot t. \qquad (3.9)$$

3.2. Model Description

During the injection process, air is entrained into the evaporated fuel zone according to a fixed relative air/fuel ratio:

$$\Lambda = 0.9, \tag{3.10}$$

and when the injection is over by means of a diffusion process:

$$\frac{dm_{vap,zone}}{d\varphi} = \frac{1}{\omega_e} \cdot c_1 \cdot Re^{c_2}(\varphi) \cdot A_{zone}(\varphi) \cdot \frac{\rho_{f,zone}(\varphi)}{d_{zone}(\varphi)}. \tag{3.11}$$

Now the fuel-mixture zone is ready for the combustion. The ignition delay comprehends a physical and a chemical part:

$$\tau_{ID}(\varphi) = \tau_{phys}(\varphi) + \tau_{chem}(\varphi), \tag{3.12}$$

and according to [6] the condition for ignition is defined as:

$$\int_{SOI}^{SOC} \frac{1}{\tau_{ID}(\varphi) \cdot \omega_e} d\varphi \geq 1. \tag{3.13}$$

Immediately after the start of combustion (SOC), the fuel-mixture of one single zone begins to burn in the premixed mode:

$$\frac{dm_{f,zone}}{d\varphi} = \min \left\{ \frac{dm_{f,zone,1}}{d\varphi}, \frac{dm_{f,zone,2}}{d\varphi} \right\}. \tag{3.14}$$

Thus at the beginning, the premixed combustion is described by:

$$\frac{dm_{f,zone,1}}{d\varphi} = \frac{1}{1 + \lambda_{zone}(\varphi) \cdot \chi_{st} \cdot (1 + r_{zone}(\varphi))} \cdot \frac{dm_b}{d\varphi} \tag{3.15}$$

$$\frac{dm_b}{d\varphi} = \frac{1}{\omega_e} \cdot \rho_u(\varphi) \cdot s_{turb}(\varphi) \cdot A_{flame}(\varphi), \tag{3.16}$$

and at the end by:

$$\frac{dm_{f,zone,2}}{d\varphi} = c_{pre} \cdot K_{pre}(\varphi) \cdot \frac{1}{\omega_e} \cdot \frac{1}{\tau_{pre}(\varphi)} \cdot m_{f,avail,zone}(\varphi), \tag{3.17}$$

where K_{pre} is a correction function depending on the amount of air, exhaust gas, and fuel in the premixed zone. The characteristic premixed time is:

$$\tau_{pre}(\varphi) = \frac{l_{zone}(\varphi)}{s_{turb}(\varphi)}, \qquad (3.18)$$

where the term s_{turb} is the turbulent flame speed.

The subsequent diffusion combustion is modeled using the so-called "mixing frequency approach." The combustion rate is proportional to the actual available evaporated fuel mass:

$$\frac{dm_{f,diff}}{d\varphi} = c_{diff} \cdot \frac{1}{\omega_e} \cdot \frac{1}{\tau_{diff}(\varphi)} \cdot m_{f,avail}(\varphi), \qquad (3.19)$$

where the available and the evaporated fuel mass during diffusion are:

$$m_{f,avail}(\varphi) = m_{vap,diff}(\varphi) - m_{f,diff}(\varphi) \qquad (3.20)$$
$$m_{vap,diff}(\varphi) = m_{vap,tot}(\varphi) - m_{vap,pre}(\varphi). \qquad (3.21)$$

The characteristic diffusion time is:

$$\tau_{diff}(\varphi) = \frac{l_{diff}(\varphi)}{u'(\varphi)}, \qquad (3.22)$$

where the term $u'(\varphi)$ describes the turbulence intensity generated by both the piston motion and the injection process.

To link both combustion modes, the diffusion combustion is multiplied with an empirical factor in order to be shifted and overlapped with the premixed combustion:

$$F_{pre/diff}(\varphi) = \left(\frac{m_{f,pre}(\varphi)}{m_{f,avail,pre}(\varphi)}\right)^{c_{pre/diff}}. \qquad (3.23)$$

The resulting evaporated, available, and burned fuel mass for both the premixed and the diffusion combustion fractions are shown exemplarily in Fig. 3.3 for the static operating point 1750 rpm and 2 bar BMEP.

3.2. Model Description

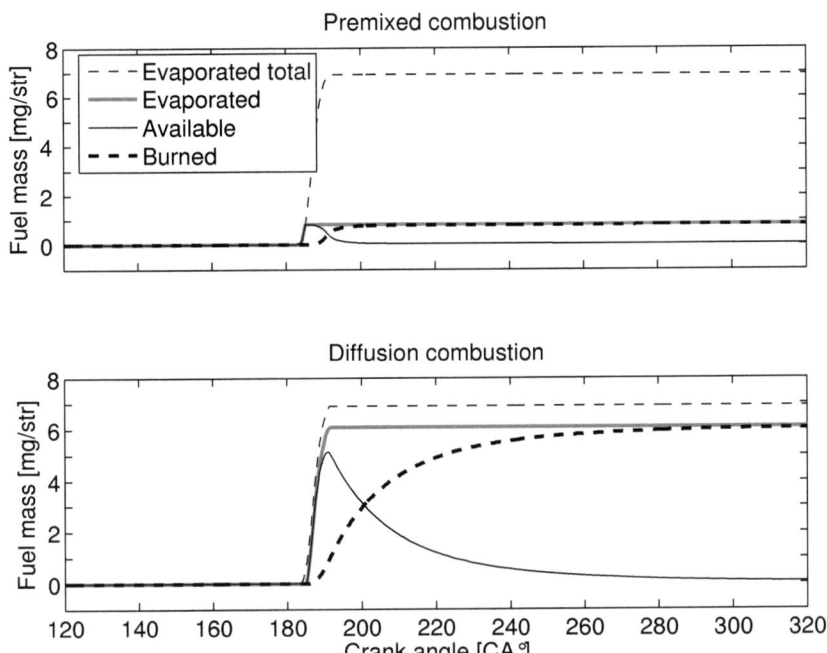

Figure 3.3: Evaporated, available, and burned fuel mass for both the premixed and the diffusion combustion fractions for the static operating point 1750 rpm and 2 bar BMEP.

3.2.3 Single-Zone Engine Process

This sub-model represents the core of the entire combustion model, because here all the in-cylinder variables such as pressure, mass, mean temperature, and λ are computed by solving energy and mass balance equations.

The inputs for this sub-model are air mass flow $\dot{m}_{air}$, inlet pressure p_{in} and temperature T_{in}, outlet pressure p_{out}, engine cooling water temperature T_e, and intake port shut-off signal u_{IPSO}. The in-cylinder gas is modelled as a single-zone medium, i.e., at each moment the in-cylinder variables are assumed to be average values valid over the whole cylinder volume V_{cyl}.

For the complete set of equations and parameters of this sub-model, refer to Appendix A.3.

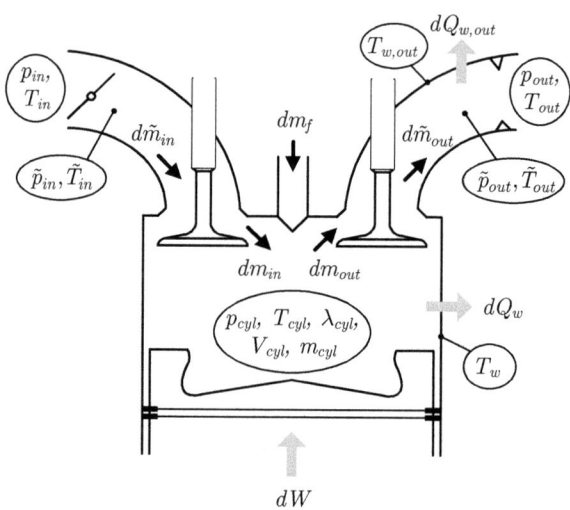

Figure 3.4: Schematic diagram of the single-zone cylinder process.

By neglecting the cylinder blow-by effects, the mass conservation is:

$$\frac{dm_{cyl}}{d\varphi} = \frac{dm_f}{d\varphi} + \frac{dm_{in}}{d\varphi} + \frac{dm_{out}}{d\varphi}, \quad (3.24)$$

and the energy conservation is:

$$\frac{dU}{d\varphi} = \frac{dQ_f}{d\varphi} + \frac{dQ_w}{d\varphi} + \frac{dW}{d\varphi} + h_{in} \cdot \frac{dm_{in}}{d\varphi} + h_{out} \cdot \frac{dm_{out}}{d\varphi}, \quad (3.25)$$

where the rates of heat release and work are:

$$\frac{dQ_f}{d\varphi} = H_f \cdot \frac{dm_f}{d\varphi} \quad (3.26)$$

$$\frac{dW}{d\varphi} = -p_{cyl}(\varphi) \cdot \frac{dV_{cyl}}{d\varphi}, \quad (3.27)$$

3.2. Model Description

respectively. The relative A/F ratio in the cylinder during the combustion process is computed on the basis of the actual air and burned fuel masses:

$$\lambda_{cyl}(\varphi) = \frac{m_{cyl,air}(\varphi)}{\chi_{st} \cdot m_{cyl,f}(\varphi)}, \quad (3.28)$$

and the corresponding global value is defined as the value at EVO, i.e., when the combustion process is over:

$$\lambda_{gl} = \lambda_{cyl}(\varphi = \varphi_{EVO}). \quad (3.29)$$

Cylinder Wall Heat Losses

For the computation of the cylinder wall heat losses, only the gas convection is taken into account:

$$\frac{dQ_w}{d\varphi} = \frac{1}{\omega_e} \cdot A_w(\varphi) \cdot q_{w,gc}(\varphi), \quad (3.30)$$

where the convective heat flow $q_{w,gc}$ is:

$$q_{w,gc}(\varphi) = \alpha(\varphi) \cdot (T_{cyl}(\varphi) - T_w(T_e)). \quad (3.31)$$

The convective heat transfer coefficient α is calculated using an extended Woschni's approach according to [16]:

$$\alpha(\varphi) = 127.93 \cdot d_{bore}^{-0.2} \cdot p_{cyl}^{0.8}(\varphi) \cdot T_{cyl}^{-0.53}(\varphi) \cdot \nu^{0.8}(\varphi). \quad (3.32)$$

The term ν describes the turbulence variation during the whole cylinder process. Two different formulation are implemented for the high pressure cycle and for the gas exchange phase. The necessary cylinder wall temperature T_w is estimated as a function of the cooling water temperature T_e with the following empirical model:

$$T_w(T_e) = T_w^0 \cdot \left(\frac{T_e}{T_e^0}\right)^c, \quad (3.33)$$

where T_w^0 and T_e^0 are reference values.

Air and Exhaust Gas Properties

The specific enthalpy of both air and exhaust gas is calculated using an approach similar to [6]:

$$h(T) = R \cdot T \cdot \left(a_0 + a_1 \cdot T + a_2 \cdot T^2 + a_3 \cdot T^3 + a_4 \cdot T^4\right), \quad (3.34)$$

where the coefficients a_i are stored in tables for the various components. Other approach are also possible, for instance according to Justy or to Zacharias, see [16]. The specific internal energy is:

$$u(T) = h(T) - R \cdot T, \quad (3.35)$$

and the specific heat at constant pressure and volume are:

$$c_p(T) = \frac{\partial h}{\partial T} \quad (3.36)$$

$$c_v(T) = \frac{\partial u}{\partial T} = c_p(T) - R, \quad (3.37)$$

respectively. The ratio of the specific heats is defined as the isentropic exponent:

$$\kappa(T) = \frac{c_p(T)}{c_v(T)}. \quad (3.38)$$

Exhaust Gas Recirculation

The EGR rate is defined as the ratio between the recirculated mass flow and the total mass flows through the cylinders:

$$X_{EGR} = \frac{\dot{m}_{EGR}}{\dot{m}_{inj} + \dot{m}_{air} + \dot{m}_{EGR}}. \quad (3.39)$$

The air mass flow and the fuel mass flow can be easily determined with ECU measurements. The recirculated gas mass flow is defined as the difference between the real gas mass flow entering the cylinders and the air mass flow:

$$\dot{m}_{EGR} = \dot{m}_{in} - \dot{m}_{air}. \quad (3.40)$$

3.2. Model Description

To compute the real gas mass flow entering the cylinders, first the theoretical value is determined on the basis of the gas equation:

$$\dot{m}_{in,th} = \frac{p_{in} \cdot (V_d \cdot n_{cyl})}{R \cdot T_{in}} \cdot \frac{\omega_e}{4\pi}, \qquad (3.41)$$

and then the theoretical value is multiplied with the volumetric efficiency in order to capture the effects of the real pumping characteristics of the engine:

$$\dot{m}_{in} = \dot{m}_{in,th} \cdot \eta_{vol}. \qquad (3.42)$$

The volumetric efficiency is defined by a pressure dependent part $\eta_{vol,p}$, an engine speed dependent part $\eta_{vol,\omega}$, and a correction term $\Delta\eta_{vol,T}$ due to intake temperature. It is a slightly different approach than the one used in [4]:

$$\eta_{vol} = \eta_{vol,p}(p_{in}, p_{out}) \cdot \eta_{vol,\omega}(\omega_e) + \\ + \Delta\eta_{vol,T}(T_{ic}, T_{in}). \qquad (3.43)$$

Gas Exchange

For the gas exchange process, the in- and outflowing masses are computed modeling the valves as isenthalpic throttles:

$$\frac{dm}{d\varphi} = \frac{1}{\omega_e} \cdot \mu\sigma \cdot A_v \cdot \frac{p_0}{\sqrt{R \cdot T_0}} \cdot \sqrt{\frac{2\kappa}{\kappa - 1} \left(\Pi_v^{\frac{2}{\kappa}} - \Pi_v^{\frac{\kappa+1}{\kappa}} \right)}, \qquad (3.44)$$

where the subscript "0" indicates the states upstream, A_v the variable valve opening area, and $\mu\sigma$ the variable discharge coefficient. The pressure ratio across the valve is defined as:

$$\Pi_v = \min \left\{ \max \left[\frac{p}{p_0}, \frac{2}{\kappa+1}^{\frac{\kappa}{\kappa-1}} \right], 1 \right\}. \qquad (3.45)$$

The test-bench engine is provided with an intake port shut-off (IPSO) system in order to decrease the PM emission at low engine speed and load, where the turbulence intensity is too low for an adequate soot oxidation process. In such operating points, one intake port (filling port) of each cylinder is closed by means of special throttles. Consequently the whole air mass is forced to flow only through the second intake port (swirl port), and thus the air enters the cylinders with increased turbulence, enhancing the soot oxidation. To capture the effect of the IPSO on the filling port, additional variables for the pressure $\tilde{p}$ and the temperature $\tilde{T}$ between the intake valve and the IPSO throttle are introduced. A similar approach is also used to model the exhaust duct, according to Fig. 3.4. The IPSO throttle reduces the cross-section of the filling port according to the following proportional coefficient:

$$c_{IPSO} = 1 - \cos\left(u_{IPSO} \cdot \frac{\pi}{2}\right). \tag{3.46}$$

The intake and exhaust port wall heat losses are also taken into account according to [22]. The variation of the air and the burned fuel mass during the gas exchange phase are computed as:

$$\frac{dm_{air,gex}}{d\varphi} = \frac{dm_{in}}{d\varphi} \cdot x_{in} + \frac{dm_{out}}{d\varphi} \cdot x_{cyl}(\varphi) \tag{3.47}$$

$$\frac{dm_{f,gex}}{d\varphi} = \frac{dm_{in}}{d\varphi} \cdot (1 - x_{in}) + \frac{dm_{out}}{d\varphi} \cdot (1 - x_{cyl}(\varphi)), \tag{3.48}$$

respectively. The air mass fraction in the intake manifold x_{in} is computed according to [30] as a function of the actual EGR rate and the actual global relative A/F ratio, whereas the air mass fraction in the cylinder x_{cyl} is found on the basis of the actual air and cylinder masses:

$$x_{in} = 1 - \frac{X_{EGR}}{1 + \lambda_{gl} \cdot X_{st}} \tag{3.49}$$

$$x_{cyl}(\varphi) = \frac{m_{air}(\varphi)}{m_{cyl}(\varphi)}. \tag{3.50}$$

3.2.4 Reaction Temperature

Since the final NO_x emission model requires information about the temperature profile during the combustion reaction, the computed cylinder mean temperature is not sufficient to perform an accurate

3.2. Model Description

computation. The in-cylinder gas is modeled as a single-zone medium, thus a method is applied to virtually separate this single zone in two distinct regions, one containing the burned and the other the unburned gases. The two zones are ideally separated by an infinitesimally thin flame front, and each zone is characterized by its own temperature profile.

For this purpose the approach proposed in [30] and [48] is used, where the temperature difference between the two virtual gas zones (burned and unburned) is described by means of the following equation:

$$T_b(\varphi) - T_u(\varphi) = A^*(\lambda_{gl}) \cdot B(\varphi), \quad (3.51)$$

where A^* denotes the maximum temperature difference between burned and unburned zone at SOC, and B is an empirical function which can vary only between 0 and 1 and that describes the temporal progression of the temperature difference. This temporal progression is governed by the energy transfer from the burned to the unburned zone, see [19], and the rate of this transfer is empirically described by means of the pressure difference between fired and motored engine operation. A similar approach is also used to model the turbulence increase due to combustion in Woschni's wall heat transfer model:

$$A^*(\lambda_{gl}) = A \cdot \frac{1.2 + (\lambda_{gl} - 1.2)^C}{2.2 \cdot \lambda_r} \quad (3.52)$$

$$B(\varphi) = \frac{K - \int_{SOC}^{\varphi} (p_{cyl}(\varphi) - \tilde{p}_{cyl}(\varphi)) \cdot m_b(\varphi) \cdot d\varphi}{K} \quad (3.53)$$

$$K = \int_{SOC}^{EVO} (p_{cyl}(\varphi) - \tilde{p}_{cyl}(\varphi)) \cdot m_b(\varphi) \cdot d\varphi. \quad (3.54)$$

The parameter A is an engine-specific value that has to be determined experimentally. It is the only tuning parameter for this sub-model. Since the combustion reaction evolves in close to stoichiometric conditions [6], the relative A/F ratio during the reaction is assumed to be constant and equal to 1:

$$\lambda_r = 1. \quad (3.55)$$

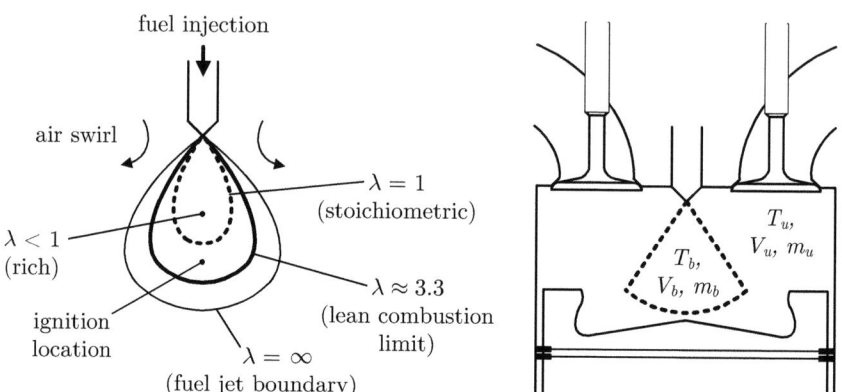

Figure 3.5: Schematic of a diesel fuel spray according to [6] (left), and the modeling of the reaction zone according to [30] and [48] (right).

For the complete set of equations and parameters of this sub-model, refer to Appendix A.4.

The advantages of this method against a real two-zones modeling are its simplicity, the reduced computational burden, and the good results. On the other hand this approach is empirical, and thus not completely based on physics.

Aside from the combustion temperature T_b other variables can be derived from this sub-model, such as the volume V_b and the density ρ_b of the burned-gas zone as well as the density ρ_u of the unburned-gas zone.

Although this approach has been developed on diesel engines, it can be successfully used for other engine types. In [81], for example, the NO_x emissions of a gas engine are computed on the basis of the reaction temperature obtained with this approach.

3.2.5 Emissions

Finally, in the sub-model "emissions" the NO_x concentration is computed. It is the sum of the NO and NO_2 produced after the combustion process (the subscript x stands for 1 and 2). Often the

3.2. Model Description

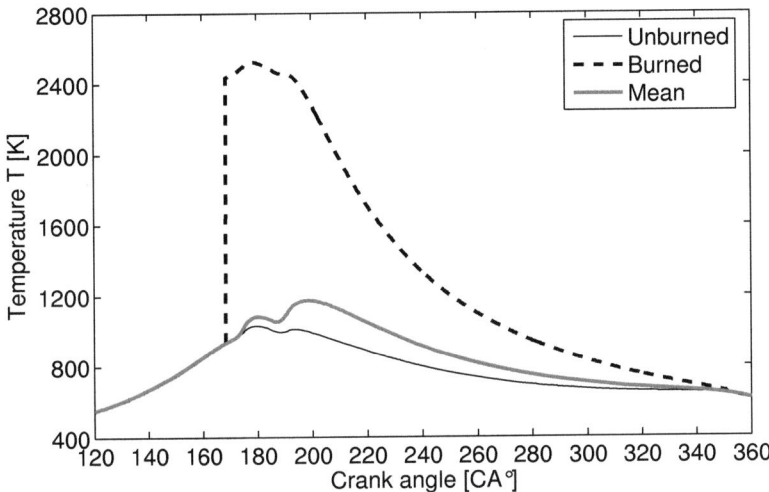

Figure 3.6: Computed characteristic temperatures during the combustion process.

modeling of the NO_2 is omitted, and only the NO are used to describe the total NO_x concentration. This can be justified, since the contribution of the NO to the total NO_x produced in combustion engines is dominant, and the NO_2 formation depends also on the amount of NO molecules present. However, here these two substances are treated separately: For the NO a standard chemical formulation is used, whereas for the NO_2 an empirical model is proposed.

For the complete set of equations and parameters of this sub-model, refer to Appendix A.5.

NO

The NO formation during an arbitrary combustion process can be described by the following four paths, see [12] and [16]:

1. thermal,
2. prompt,

3. nitrous oxide (N_2O), and

4. fuel.

In the case of the thermal NO path, the NO molecules are formed behind the flame front, i.e., in the hot burned gas zone. The name "thermal" indicates that the temperature is the main factor describing this chemical mechanism, which has been first defined by Zeldovich in 1946, and then has been extended to what nowadays is known as "extended Zeldovich" mechanism.

The prompt NO is formed under rich conditions in the flame front. The name "prompt" indicates that the reactions, which have been first described by Fenimore in 1979, are very fast. The prompt NO path is more complicated since it is directly related to the formation of the CH radical, which can react in multiple ways.

The N_2O path is relevant especially under lean conditions, low temperature, and high pressure, for instance in the case of lean premixed-combustion of gas turbines or of lean SI engine concepts.

The last path considered, i.e., the fuel-NO path, describes the formation of NO due to the nitrogen contained in the fuel. This path is not relevant for combustion engines since the amount of nitrogen contained in the fuel is negligible, but it should be considered for instance in the case of coal combustion.

Based on all this information, it is evident that only the thermal NO path plays a major role for the modeling of the NO formation in diesel engines, see also [6], [39] and [63]. It is thus the only mechanism considered here.

During a typical diesel combustion process various chemical components are formed. However, only the following twelve components have a relevance for chemical reactions: H_2O, CO, O_2, N_2, H_2, OH, CO, N_2O, NO, N, O, H. Furthermore, since the equilibrium concentrations of the components are reached more quickly than those of N and NO, it is assumed that all components except N and NO are in equilibrium. Thus, to determine the equilibrium concentrations, it is necessary to solve a set of algebraic equations, and successively, for

3.2. Model Description

the computation of the concentrations of N and NO, a set of differential equations. These differential equations are based on the extended Zeldovich mechanism, which is described as follows:

$$N_2 + O \rightleftharpoons NO + N \qquad (3.56)$$
$$O_2 + N \rightleftharpoons NO + O \qquad (3.57)$$
$$N + OH \rightleftharpoons NO + H. \qquad (3.58)$$

The final equation for the NO formation is:

$$\frac{d[NO]}{d\varphi} = \frac{1}{\omega_e} \cdot 2 \cdot \left(1 - \nu^2(\varphi)\right) \cdot \frac{R_{1e}(p_{cyl}, T_b, \lambda_r)}{1 + \nu(\varphi) \cdot K_e(p_{cyl}, T_b, \lambda_r)} - \xi(\varphi), \quad (3.59)$$

where the normalized NO concentration ν and the concentration variation ξ due to the burned zone volume variation are:

$$\nu(\varphi) = \frac{[NO](\varphi)}{[NO]_e(p_{cyl}, T_b, \lambda_r)} \qquad (3.60)$$

$$\xi(\varphi) = \frac{[NO](\varphi)}{V_b(\varphi)} \cdot \frac{dV_b}{d\varphi}. \qquad (3.61)$$

The values $[NO]_e$, R_{1e}, and K_e are calculated by solving the complete chemical mechanism for various values of pressure, temperature, and relative A/F ratio, and then stored in maps as a function of cylinder pressure, burned zone temperature, and relative A/F ratio during the reaction. This last value is assumed to be equal to 1, as shown in Eq. (3.55) above.

NO_2

According to [6] and [20], the formation of NO_2 occurs when NO molecules from high-temperature regions diffuse or are transported by fluid mixing into the HO_2-rich regions. Furthermore, HO_2 radicals are formed in a relatively low-temperature region. The NO_2 destruction reactions are active at high temperatures, thus preventing the formation of NO_2 in high temperature zones. For the NO_2 formation the following equation holds:

$$NO + HO_2 \rightleftharpoons NO_2 + OH, \qquad (3.62)$$

and for the NO_2 destruction:

$$NO_2 + H \rightleftharpoons NO + OH \qquad (3.63)$$
$$NO_2 + O \rightleftharpoons NO + O_2, \qquad (3.64)$$

where:

$$H + O_2 + M \rightleftharpoons HO_2 + M. \qquad (3.65)$$

Since the chemical mechanisms involved in the NO_2 formation are quite complicated, and the contribution of NO_2 to the total NO_x concentration is not of a main relevance, as the contribution of NO is, a simple empirical model for the NO_2/NO concentrations ratio is proposed:

$$\frac{[NO_2]}{[NO]} = c_1 + c_2 \cdot [NO]^{-1}. \qquad (3.66)$$

3.3 Model Calibration

Once all geometrical data of the engine and its components are collected, static measurements are performed in order to optimize the parameters of the physical model. The fuel injection sub-model is calibrated first. A newly developed tool evaluates the static injection measurements and minimizes the difference between the injected fuel quantities measured and calculated in order to find the optimal parameters. This is done on the basis of the Nelder-Mead algorithm. This procedure is followed for the pilot injection as well as for the main injection in order to parameterize both injection parts separately.

Successively, the heat release sub-model is calibrated by means of data obtained from a cylinder pressure analysis. Cylinder pressure traces for one cylinder are recorded by means of a non-cooled cylinder pressure sensor placed in the glow plug hole, and processed in

3.3. Model Calibration

order to obtain a virtual measurement of the heat release rate with a cylinder pressure analysis tool developed at the Aerothermochemistry and Combustion Systems Laboratory (LAV) of the ETH Zurich named "WEG," see Fig. 3.7. This procedure is repeated for various static operating points. The heat release rate traces obtained serve to optimize manually the parameters of the equations describing the combustion and the ignition delay processes. Since the model is physics-based and thus the different parameters have a physical meaning, only small adjustments of the parameters are necessary. Furthermore, the number of parameters to be adjusted is limited to a few fundamental quantities, such that the manual optimization is a quick and intuitive process. Some calibration results are shown in Figs. 3.8-3.10 below.

Next, the EGR calculation sub-model is calibrated. Another newly developed tool evaluates the static EGR measurements and computes the parameters for the volumetric efficiency on the basis of a regression analysis.

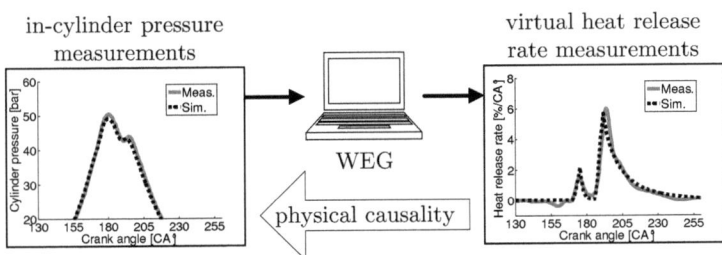

Figure 3.7: The measured cylinder pressure signals are evaluated by means of "WEG", a tool by ETH-LAV, in order to find the corresponding heat release rate.

Finally, the reaction temperature sub-model is calibrated. A third specifically developed tool evaluates the static NO_x concentration measurements and minimizes the difference between measured and calculated NO_x concentrations in order to find the optimal engine-specific A-parameter of Eq. (3.52) for all operating points considered, again on the basis of the Nelder-Mead algorithm.

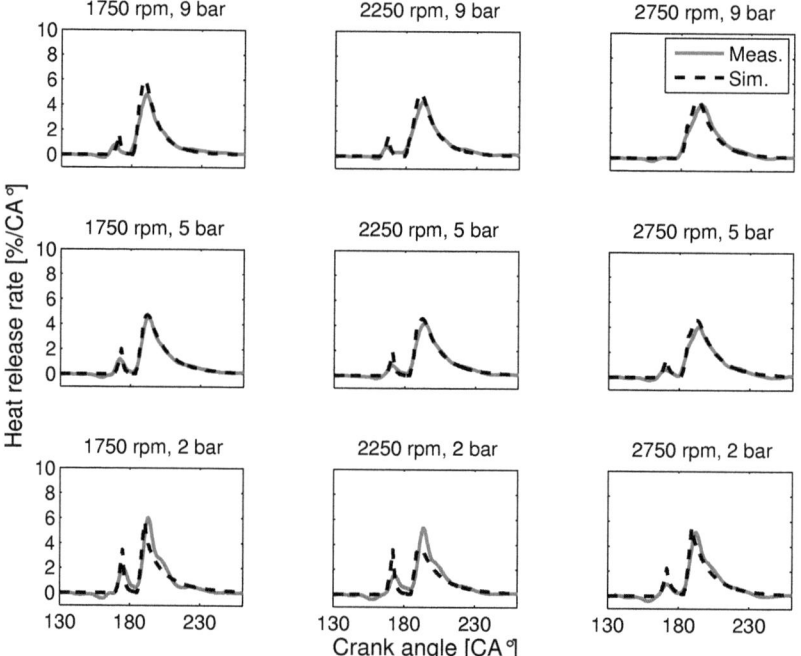

Figure 3.8: Calibration of the combustion model: Measured vs. simulated heat release rate for various static operating points.

3.3. Model Calibration

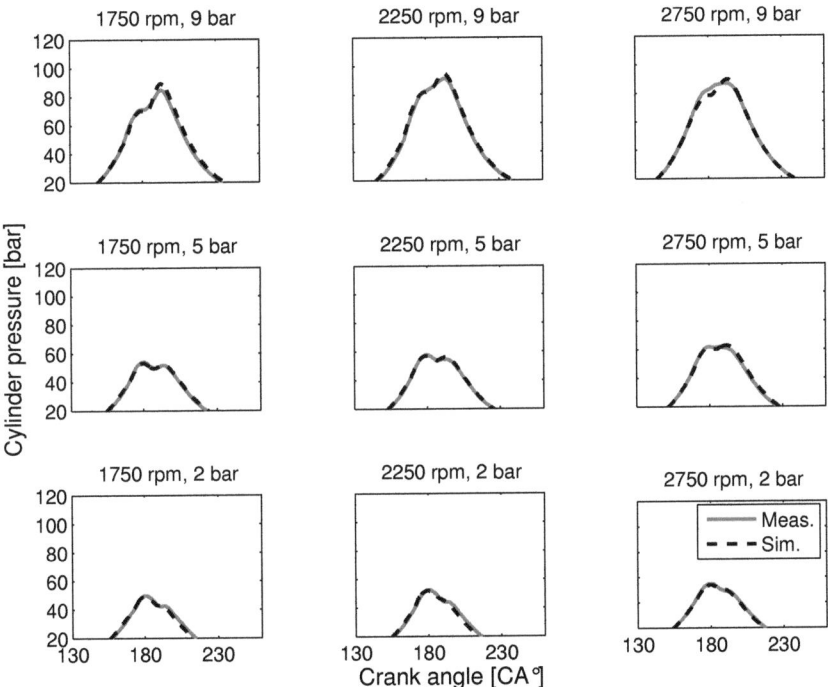

Figure 3.9: Calibration of the combustion model: Measured vs. simulated cylinder pressure for various static operating points.

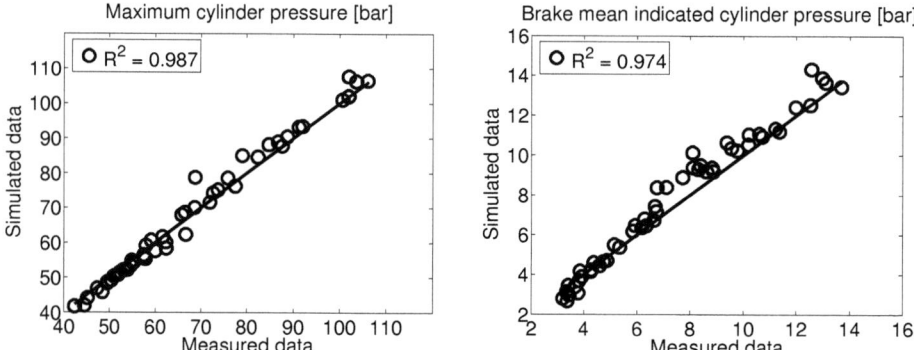

Figure 3.10: Calibration of the combustion model: Measured vs. simulated maximum cylinder pressure (left), and indicated mean cylinder pressure (right).

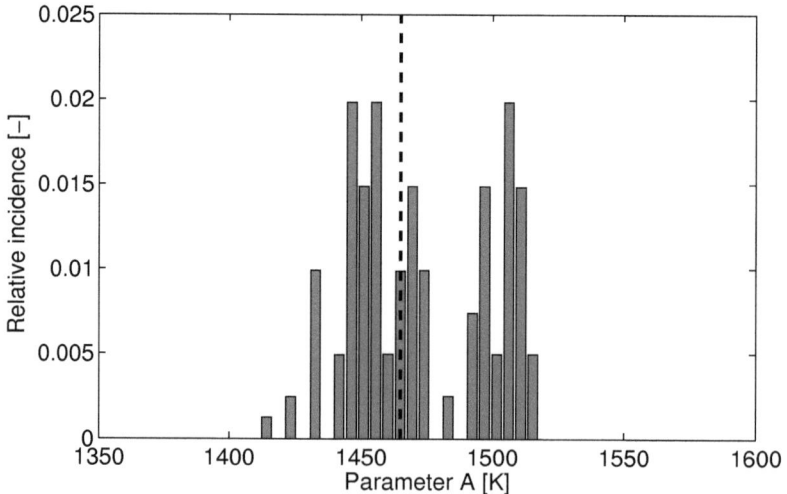

Figure 3.11: Calibration of the combustion model: Finding the optimal engine-specific A-parameter.

3.4. Model Verification

The optimal A-parameter is determined on the basis of the relative incidence of all A-values found, as shown in Fig. 3.11.

The static measurements used to calibrate the combustion model are those shown in Fig. 3.12 left. For the cylinder pressure analysis with WEG, different measurement data are used than for the other sub-models.

3.4 Model Verification

The model prediction capabilities are verified on the basis of other static measurements than those used to calibrate the model. They are shown in Fig. 3.12 right.

Verification results are shown below. Figure 3.13 shows the comparison between measured and simulated injected fuel quantity during pilot (left) and main injection (right). Figure 3.14 shows the comparison between measured and simulated NO_2 concentration as well as the corresponding NO_x levels. In this case, the NO_2 and NO_x measurement data are obtained with the CLD-device. Figure 3.15 shows the comparison between measured and simulated EGR rate (left) and NO_x concentration (right). In this case, the NO_x measurement data are obtained with the solid-state sensor.

To further verify the prediction capabilities of the combustion model, Figs. 3.16-3.19 show the comparisons between measured and simulated NO_x concentration for variations of the engine inputs "start of injection," "rail pressure," "air mass flow," and "boost pressure" for four different operating points. In these cases too the NO_x measurement data are obtained with the solid-state sensor.

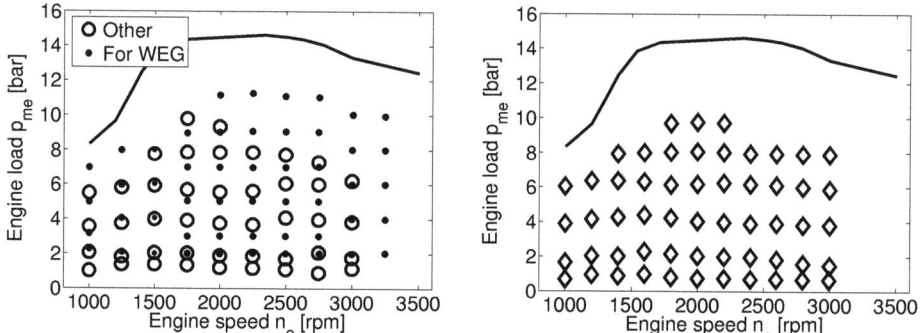

Figure 3.12: Operating points used for the calibration (left) and the verification (right) of the combustion model.

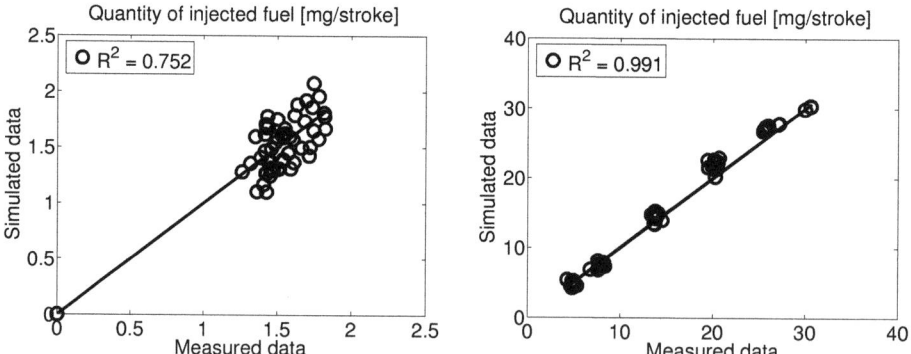

Figure 3.13: Verification of the combustion model: Measured vs. simulated quantity of injected fuel for pilot injection (left), and main injection (right).

3.4. Model Verification

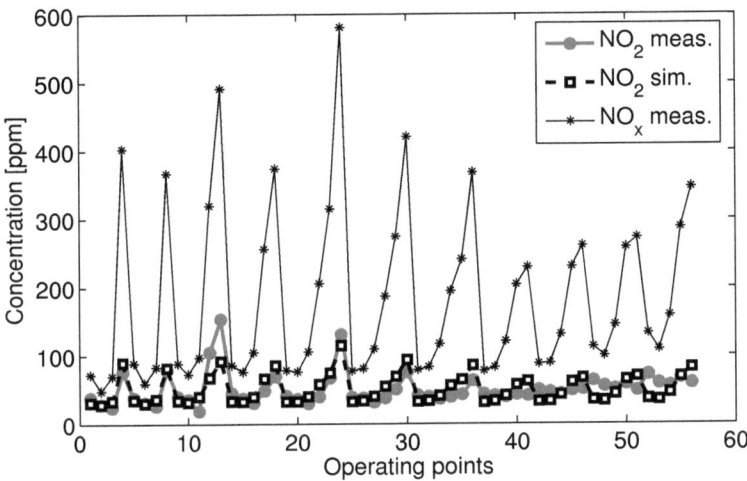

Figure 3.14: Verification of the combustion model: Measured vs. simulated NO_2 concentration for different static operating points.

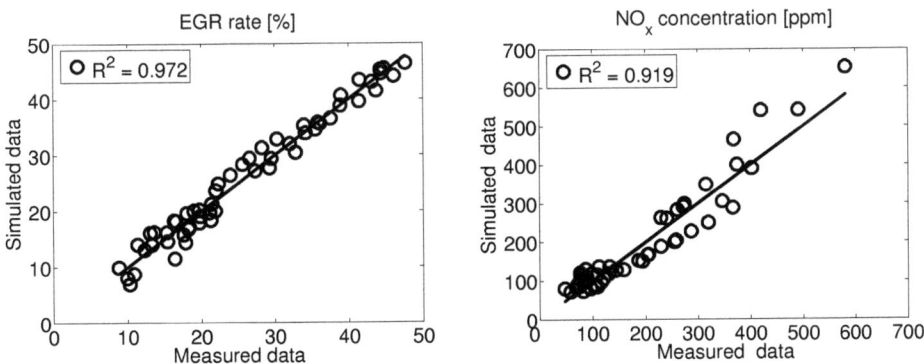

Figure 3.15: Verification of the combustion model: Measured vs. simulated EGR rate (left) and NO_x concentration (right).

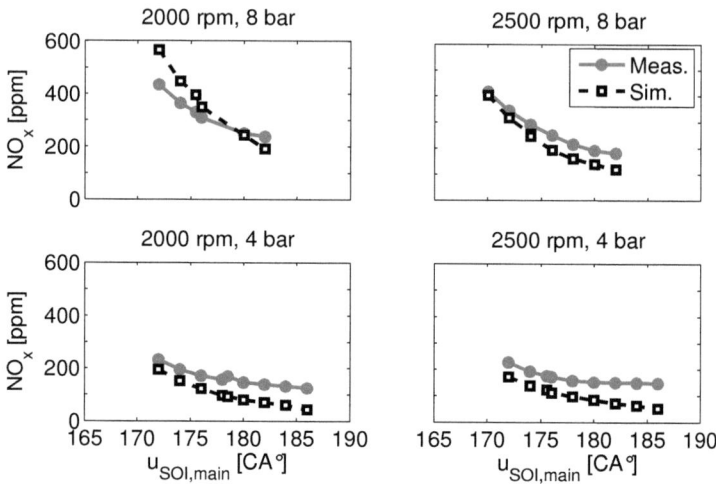

Figure 3.16: Verification of the combustion model: Measured vs. simulated NO_x for variations of start of injection at static operating points.

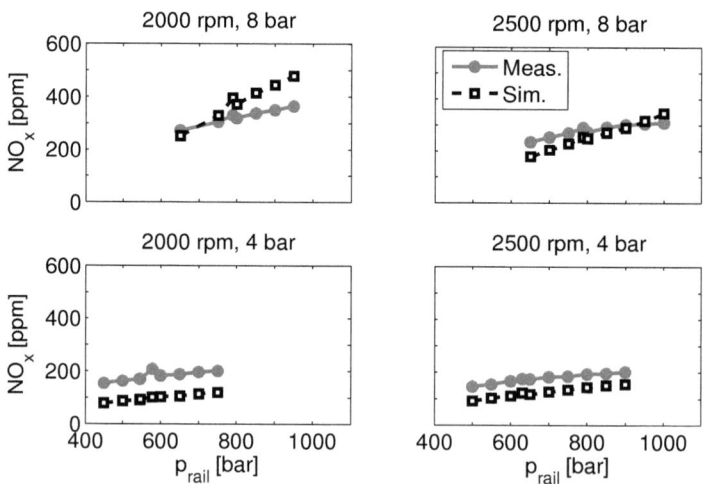

Figure 3.17: Verification of the combustion model: Measured vs. simulated NO_x for variations of rail pressure at static operating points.

3.4. Model Verification

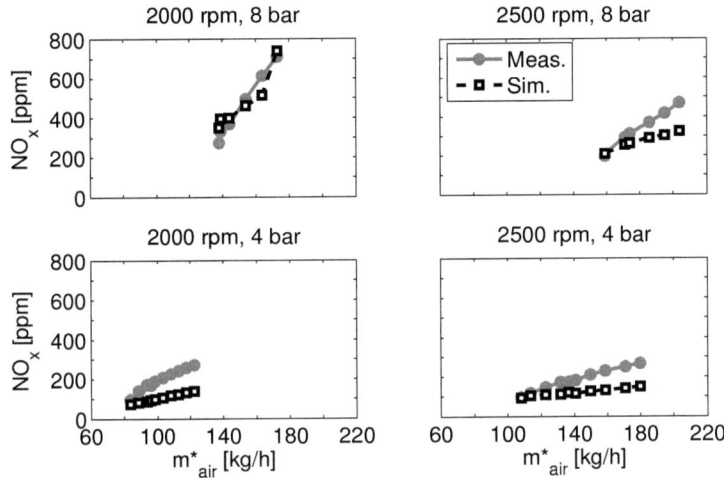

Figure 3.18: Verification of the combustion model: Measured vs. simulated NO_x for variations of air mass flow at static operating points.

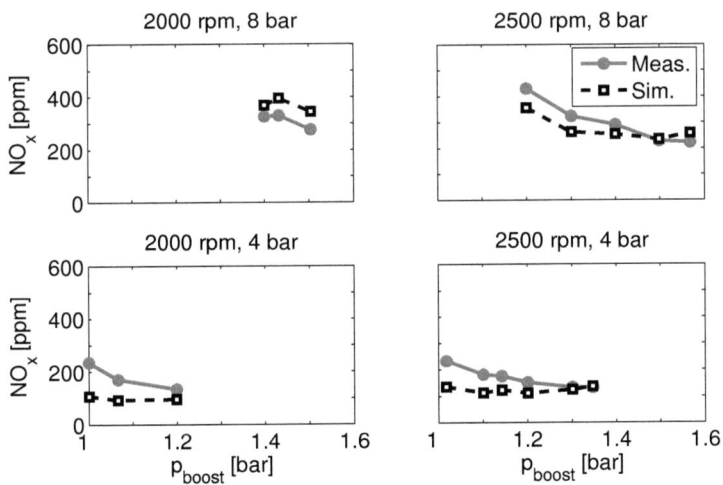

Figure 3.19: Verification of the combustion model: Measured vs. simulated NO_x for variation of boost pressure at static operating points.

Comments on the Results

The capabilities of the combustion model developed are reassumed as follows:

- the maximum relative error of the whole model at low NO_x concentrations is about 40 %, whereas at high NO_x concentrations is less than about 20%. This indicates that the model performances are better at high NO_x levels than at low;

- the model is in general able to capture the trends resulting from variations of the engine inputs. However, in the case of the variations of the air mass flow, the model clearly underestimated the NO_x concentrations;

- the proposed empirical NO_2 model of Eq. (3.66) gives good results.

The differences between the model and the measurement can be explained as follows: The model is conceived to calculate the concentration of the NO_x using injection parameters and air path measurements directly. According to Fig. 3.1, the model comprehends the main processes involved during the combustion, and the computation starts from information about the fuel injection and the breathing of the engine. Hence the modeling errors of all five sub-models are cumulated during the computation.

Chapter 4
Control-Oriented Emission Models

Based on the detailed combustion model developed, two models are presented for the real-time computation of the λ value and the NO_x emission. These models should rely only on measurement signals available from the ECU of the engine and must be sufficiently accurate. Furthermore, an adaptive strategy is implemented for the NO_x model in order to enhance its accuracy, and to deal with engine variability as well as with modeling uncertainties.

4.1 Motivation and Previous Work

Since the FDI system proposed in this work is model-based, appropriate models for the prediction of the λ and NO_x levels are needed. These models should be:

- based on available ECU measurements,
- applicable for real-time purposes,
- accurate, and

- easily transferable to other engines.

The computation of λ is straightforward, since according to its physical definition it is directly proportional to the air quantity and inversely proportional to the fuel quantity. Unfortunately, the NO_x emission formation is a more complicated process with more variables involved, and thus the derivation of a simplified formulation for the NO_x model needs more detailed approaches.

Various methods to obtain a real-time NO_x model have been proposed in the literature. A simple one has been presented in [35] and [112], where the NO_x model consists of a map of the NO_x emission obtained from static measurements taken at different engine operating points. This approach is straightforward to implement, but, obviously, requires many experiments. Moreover, the accuracy is not acceptable for this application, since the effects of any engine input variations from the conditions where the map measurements have been conducted are not taken into account.

An interesting physics-based approach has been proposed in [117], where a detailed chemical NO_x model has been simplified in order to obtain a reduced formulation for the computation of the averaged NO_x concentration over each combustion cycle. This reduced formulation contains an empirical function to be defined by the user that describes the rate of NO_x production with respect to the various engine inputs. The parameters of this user-defined function are found by means of an optimization routine. Although this approach seems to be powerful, the form of the empirical function for the rate of NO_x production strongly influences the quality of the model. Furthermore, the choice of an adequate function is not trivial, and no methodology to define it is given.

In [106] another physics-based model for HiL purposes is described. The resulting model is composed of a zero-dimensional formulation of the diesel combustion, the estimation of the combustion temperature, the determination of the chemical equilibrium concentrations, and the NO formation based on the Zeldovich mechanism. The model is promising for RT purposes, but its accuracy is limited in the range

4.1. Motivation and Previous Work

of common operating conditions of diesel engines.

Even black-box approaches have been investigated. In [91] a method based on genetic algorithms able to determine the structure and the parameters of a formula for the NO_x calculation is presented. The identification has been performed for measurements with EGR as well as without EGR, yielding two completely different formulas for the NO_x emission. Although in both cases the results shown are excellent, the reference to physics is completely missing, and therefore the NO_x prediction is valid only under operating conditions similar to those used for the model identification.

In [47] a fast neural network (NN) with local linear models and normalized radial basis functions is proposed. Those models are able to capture the static as well as the dynamic behavior of the concentrations of exhaust gas components such as NO_x and soot. However, their application for control purposes is still not possible in actual ECUs because of the considerable calculation effort needed. Furthermore, the amount of measured data needed for the training of the NN is considerable because it has to contain the whole range of amplitudes and dynamic effects of the process. Accordingly, the application of an NN to other engines is not straightforward, and the training procedure has to be repeated every time.

In [100] and [102] a hybrid solution between the classical physics-based methods and the black-box approach is proposed. The NN are used only where no physical descriptions of the processes are available, where the accuracy of the physical modeling is poor, or where the use of physics-based formulations would imply a considerable computational time. Although this solution seems to be promising, the presence of NN implies the same drawbacks as those described above.

In [116] the signals obtained from cylinder pressure sensors are evaluated in order to obtain combustion relevant information, such as combustion duration, location of the 50% mass fraction burned, and maximum burning rate. This information then provides the inputs for an NN, since it strongly correlate with emissions. Again, since it is based on NN, this approach is also faced with the problems described above.

4.2 Virtual Sensors for λ and NOx

However, since these methods are not able to satisfy all the conditions listed above, for the real-time estimation of both λ and NO_x, control-oriented models following the guidelines presented in [113] are developed. The details of this approach are explained in the following section.

4.2.1 Description of the Models

The reference values λ^* and NO_x^* of the actual λ and NO_x are corrected by means of the corresponding normalized variation $\delta\lambda$ and δNO_x according to Eqs. (4.1) and (4.2):

$$\lambda(k) = \lambda^*(k) \cdot [1 + \delta\lambda(k)] \quad (4.1)$$
$$NO_x(k) = NO_x^*(k) \cdot [1 + \delta NO_x(k)] \quad (4.2)$$

The computation of the normalized variations $\delta\lambda$ and δNO_x of the actual measured values of λ and NO_x from their reference values is performed by means of a linear combination of the normalized variations of the actual engine inputs from their reference values according to Eqs. (4.3) and (4.4). The factors θ_{ij} in these equations are the sensitivities of the engine output y_j (λ and NO_x) with respect to each single engine input w_i considered:

$$\begin{aligned}\delta\lambda(k) = {} & \delta\tau_{inj}(k) \cdot \theta_{1\lambda}(k) + \delta p_{rail}(k) \cdot \theta_{2\lambda}(k) + \\ & + \delta\dot{m}_{air}(k) \cdot \theta_{3\lambda}(k)\end{aligned} \quad (4.3)$$

$$\begin{aligned}\delta NO_x(k) = {} & \delta u_{SOI}(k) \cdot \theta_{1N}(k) + \delta\tau_{inj}(k) \cdot \theta_{2N}(k) + \\ & + \delta p_{rail}(k) \cdot \theta_{3N}(k) + \delta\dot{m}_{air}(k) \cdot \theta_{4N}(k) + \\ & + \delta p_{boost}(k) \cdot \theta_{5N}(k) + \delta T_e(k) \cdot \theta_{6N}(k).\end{aligned} \quad (4.4)$$

The models are completed with the definition of the normalized variations δw_i of each measured engine input w_i^m from its reference value

4.2. Virtual Sensors for λ and NOx

w_i^*, which is:

$$\delta w_i(k) = \frac{w_i^m(k) - w_i^*(k)}{w_i^*(k)}. \tag{4.5}$$

The parameters θ_{ij} for the linear combinations of Eqs. (4.3) and (4.4), which have the meaning of sensitivities, are listed in Table 4.1. They are obtained from the detailed combustion model presented above, and their computation is described in the next section.

Table 4.1: Parameters of the virtual sensors for λ and NO_x.

	$\frac{\partial y}{\partial u_{SOI}}$	$\frac{\partial y}{\partial \tau_{inj}}$	$\frac{\partial y}{\partial p_{rail}}$	$\frac{\partial y}{\partial \dot{m}_{air}}$	$\frac{\partial y}{\partial p_{boost}}$	$\frac{\partial y}{\partial T_e}$
$y = \lambda$	-	$\theta_{1\lambda}$	$\theta_{2\lambda}$	$\theta_{3\lambda}$	-	-
$y = NO_x$	θ_{1N}	θ_{2N}	θ_{3N}	θ_{4N}	θ_{5N}	θ_{6N}

The resulting virtual sensors are schematically shown in Fig. 4.1. The detailed combustion model mentioned above served also to choose the relevant engine inputs for the NO_x emission model based on a sensitivity analysis. This analysis showed that only the inputs "start of injection" u_{SOI}, "injection duration" τ_{inj}, "rail pressure" p_{rail}, "air mass flow" $\dot{m}_{air}$, "boost pressure" p_{boost}, and "engine cooling water temperature" T_e have a noticeable influence on the NO_x emission. Also the EGR rate has a considerable effect on the NO_x emission, but its level cannot be directly determined from the ECU measurements available. However, the EGR level is implicitly given from both the inputs "air mass flow" and "boost pressure." The choice of the engine inputs for the λ model is directly derived from its physical definition. Therefore, only the engine inputs "injection duration," "rail pressure," and "air mass flow" are taken into account[1].

According to the physical definition of λ, the choice of describing the normalized corrections of the λ values with a linear combination is

---

[1] Injection duration and rail pressure are chosen since the injected fuel quantity is controlled by these two values.

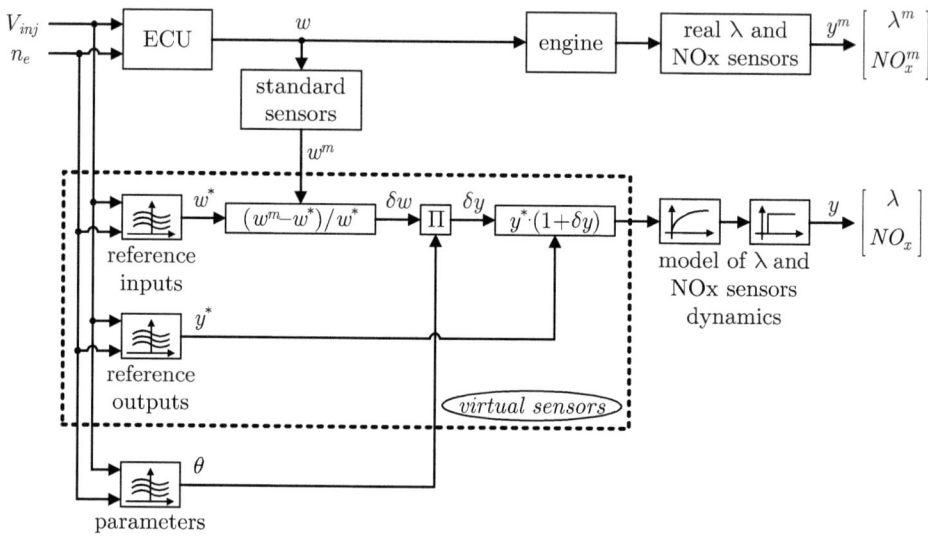

Figure 4.1: Schematic of the proposed virtual sensors for the on-line computation of λ and NO_x.

justified. In the case of the NO_x emission this choice is not straightforward, and represents a sort of approximation of the real, non-linear behavior of the NO_x emission formation in diesel engines. However, since under normal operating conditions the engine inputs do not deviate significantly from their reference values, the resulting approximation is acceptable.

Discrete State-Space Representation

To allow a comparison with the corresponding real measured signals, the virtual sensors of Eqs. (4.1)-(4.5) are completed with elements describing the characteristics of the real ones as shown in Fig. 4.1, i.e., a lag element as well as a time delay. Both λ and NO_x sensors are assumed to be first-order lag elements, whose transfer functions are

4.2. Virtual Sensors for λ and NOx

expressed by means of the discrete Laplace variable z:

$$G_\lambda(z) = \frac{b_\lambda}{z + a_\lambda}, \quad G_N(z) = \frac{b_N}{z + a_N}. \quad (4.6)$$

The discrete time delay, which consists of both the gas transport and the sensor delays, is defined as:

$$n_{d\lambda} = \text{round}\left(\frac{\tau_{d\lambda}}{T_{sample}}\right), \quad n_{dN} = \text{round}\left(\frac{\tau_{dN}}{T_{sample}}\right). \quad (4.7)$$

The discrete λ sensor parameters a_λ and b_λ, the discrete NO_x sensor parameters a_N and b_N, as well as the time delays $\tau_{d\lambda}$ and τ_{dN} are determined experimentally and saved in maps as a function of the operating conditions.

The models are completed with the input noise vector v and the measurement noise vector q. The resulting models are time varying, but linear. Therefore, they can be described by means of the following discrete state-space representation:

- discrete system equations:

$$\begin{cases} x(k+1) = \overbrace{A \cdot x(k) + B(k) \cdot u(k)}^{f(x(k),u(k))} + B_v(k) \cdot v(k) \\ y(k) = \underbrace{C \cdot x(k - n_d)}_{g(x(k),u(k))} + q(k) \end{cases} \quad (4.8)$$

- input, state, and output vectors:

$$u(k) = \begin{bmatrix} 1 \\ 1 \\ \delta u_{SOI}(k) \\ \delta \tau_{inj}(k) \\ \delta p_{rail}(k) \\ \delta \dot{m}_{air}(k) \\ \delta p_{boost}(k) \\ \delta T_e(k) \end{bmatrix}, \quad x(k) = \begin{bmatrix} \lambda(k) \\ NO_x(k) \end{bmatrix}$$

$$y(k) = \begin{bmatrix} \lambda(k - n_{d\lambda}) \\ NO_x(k - n_{dN}) \end{bmatrix}, \qquad (4.9)$$

- discrete time-varying system matrices:

$$A = \begin{bmatrix} -a_\lambda & 0 \\ 0 & -a_N \end{bmatrix}, \quad C = \begin{bmatrix} 1 & 0 \\ 0 & 1 \end{bmatrix},$$

$$B(k) = \tilde{B}^*(k) \cdot \tilde{B}_\theta(k). \qquad (4.10)$$

The input system matrix B is composed of two parts: The first one, $\tilde{B}^*$, depends on the actual reference values of λ and NO_x, and the second one, $\tilde{B}_\theta$, on the actual parameters of the λ and NO_x models:

$$\tilde{B}^*(k) = \begin{bmatrix} b_\lambda \cdot \lambda^*(k) & 0 \\ 0 & b_N \cdot NO_x^*(k) \end{bmatrix}, \qquad (4.11)$$

$$\tilde{B}_\theta(k) = \begin{bmatrix} 1 & 0 & 0 & \theta_{1\lambda}(k) & \theta_{2\lambda}(k) & \theta_{3\lambda}(k) & 0 & 0 \\ 0 & 1 & \theta_{1N}(k) & \theta_{2N}(k) & \theta_{3N}(k) & \theta_{4N}(k) & \theta_{5N}(k) & \theta_{6N}(k) \end{bmatrix}.$$

The matrix B_v is the input noise matrix, and it is chosen to be equal to the system input matrix B:

$$B_v(k) = B(k). \qquad (4.12)$$

The linear functions f and g of Eq. (4.8) are composed by two parts, one describing the λ and the other the NO_x model:

$$f(x(k), u(k)) = \begin{bmatrix} f_\lambda(\lambda(k), u(k)) \\ f_N(NO_x(k), u(k)) \end{bmatrix}, \qquad (4.13)$$

$$g(x(k), u(k)) = \begin{bmatrix} g_\lambda(\lambda(k), u(k)) \\ g_N(NO_x(k), u(k)) \end{bmatrix}. \qquad (4.14)$$

4.2. Virtual Sensors for λ and NOx

Hence expanding Eq. (4.8) one can obtain:

$$\begin{bmatrix} \lambda(k+1) \\ NO_x(k+1) \end{bmatrix} = \begin{bmatrix} -a_\lambda & 0 \\ 0 & -a_N \end{bmatrix} \begin{bmatrix} \lambda(k) \\ NO_x(k) \end{bmatrix} +$$

$$+ \begin{bmatrix} b_\lambda \lambda^*(k) & 0 & 0 & b_\lambda \lambda^*(k)\theta_{1\lambda}(k) & \cdots \\ 0 & b_N NO_x^*(k) & b_N NO_x^*(k)\theta_{1N}(k) & b_N NO_x^*(k)\theta_{2N}(k) & \cdots \end{bmatrix}$$

$$\begin{matrix} \cdots & b_\lambda \lambda^*(k)\theta_{2\lambda}(k) & b_\lambda \lambda^*(k)\theta_{3\lambda}(k) & 0 & \cdots \\ \cdots & b_N NO_x^*(k)\theta_{3N}(k) & b_N NO_x^*(k)\theta_{4N}(k) & b_N NO_x^*(k)\theta_{5N}(k) & \cdots \end{matrix}$$

$$\begin{matrix} \cdots & 0 \\ \cdots & b_N NO_x^*(k)\theta_{6N}(k) \end{matrix} \Bigg] \begin{bmatrix} 1 \\ 1 \\ \delta u_{SOI}(k) \\ \delta \tau_{inj}(k) \\ \delta p_{rail}(k) \\ \delta \dot{m}_{air}(k) \\ \delta p_{boost}(k) \\ \delta T_e(k) \end{bmatrix} + B_v(k)v(k). \quad (4.15)$$

4.2.2 Calculation of the Model Parameters

The parameters of the proposed control-oriented models are defined as the sensitivities of λ and NO_x emission levels to the corresponding relevant engine inputs. Therefore they are computed by determining the ratios of the normalized variation of the λ value and of the NO_x emission to the normalized variation of the various engine inputs considered. For example, the sensitivity θ_{ij} of the engine output y_j to the engine input w_i is defined as follows:

$$\theta_{ij} = \frac{\frac{y_j - y_j^*}{y_j^*}}{\frac{w_i - w_i^*}{w_i^*}} = \frac{\partial y_j}{\partial w_i}. \quad (4.16)$$

Thus, the sensitivities of λ and the NO_x emission for each engine input considered in each desired engine operating point have to be computed in order to derive complete maps of parameters for the control-oriented models. Intuitively, these maps of parameters can be obtained directly from measurements at the test-bench engine. However, many measurements would have to be taken, and consequently the time spent at the test-bench would be considerable. Instead, a very efficient way to obtain the necessary maps of parameters in a reduced amount of time, is to use the detailed combustion model developed in Chapter 3. It is possible to use it as a "virtual engine," capable of simulating in each desired operating point all the necessary engine input variations.

Figure 4.2 shows the method used to calculate the parameters for the case of the NO_x model: First, static measurements at the test-bench are performed in order to define the maps for the reference engine inputs and the corresponding reference NO_x emissions. Successively, the detailed combustion model is used to perform simulations of variations of different magnitudes for each single engine input in each operating condition used to define the reference maps. Then the variations of the simulated NO_x emissions from their reference values are calculated. The sensitivities are obtained by dividing the calculated variations of the NO_x emissions to the variations of each engine input. Finally, the values of the sensitivities are stored in maps as a function of the operating conditions.

Thus the sensitivities of NO_x to all engine inputs considered are computed on the basis of the simulated NO_x levels. For the sensitivities of λ an alternative approach is used, see Eqs. (4.17)-(4.19): Since λ is directly proportional to the air mass flow and inversely proportional to the injected fuel mass, its sensitivities for the engine inputs "injection duration" and "rail pressure" are computed on the basis of the simulated injected fuel mass, following the same process as shown in Fig. 4.2, and its sensitivity for the engine input "air mass flow" is set equal to one.

4.2. Virtual Sensors for λ and NOx

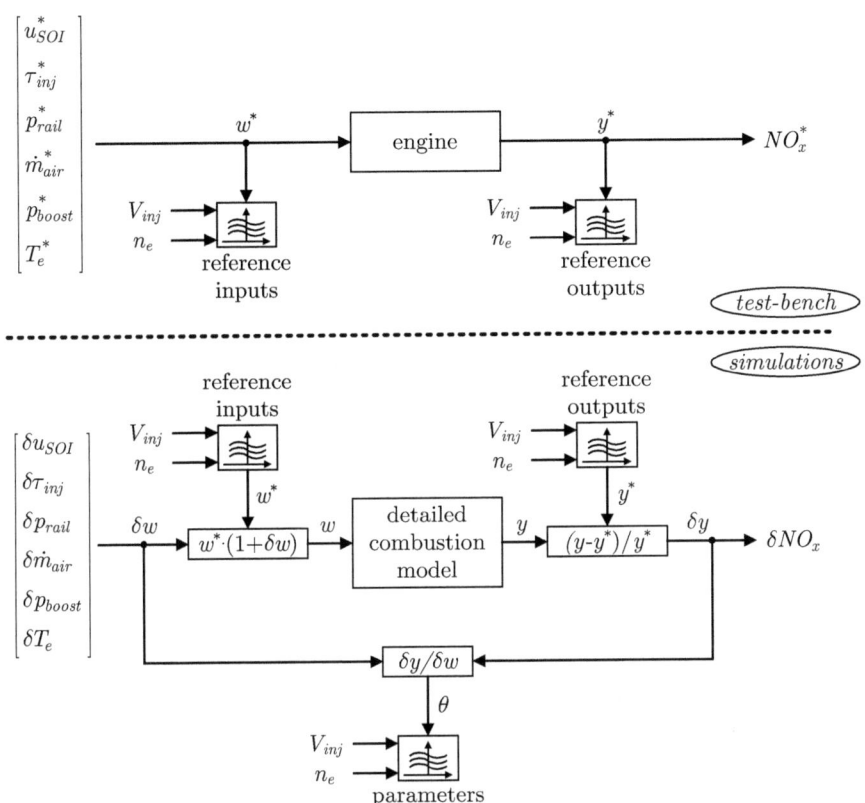

Figure 4.2: Overview of the derivation of the parameters for the NO_x model.

$$\theta_{1\lambda} = \frac{\partial \lambda}{\partial \tau_{inj}} = -\frac{\partial m_{inj}}{\partial \tau_{inj}} \tag{4.17}$$

$$\theta_{2\lambda} = \frac{\partial \lambda}{\partial p_{rail}} = -\frac{\partial m_{inj}}{\partial p_{rail}} \tag{4.18}$$

$$\theta_{3\lambda} = \frac{\partial \lambda}{\partial \dot{m}_{air}} = 1 \tag{4.19}$$

The reason for this alternative approach is that the injected fuel quantity is calculated exclusively within the injection sub-model of Fig. 3.1, which represents the first stage of the simulation of the complete combustion process. Therefore, the modeling errors are expected to be limited.

The maps of the reference input and output values are obtained on the basis of the same static operating point measurements used to perform the verification of the combustion model, according to Fig. 3.12 right.

4.2.3 Example

The following example is intended to illustrate the control-oriented models described above. The test-bench is programmed according to the engine load and speed profiles shown in Fig. 4.3. Measurement data of the various engine inputs and outputs are recorded for the standard case as well as for the case in which the EGR valve is kept closed, thus deactivating the EGR system. The normalized engine inputs, and the parameters for the λ as well as those for the NO_x models needed to solve Eqs. (4.3) and (4.4) are shown in Figs. 4.4 and 4.5, respectively, for both the cases with EGR and without EGR.

Consider the normalized engine inputs of Fig. 4.4, which ideally should be near zero. The first sub-plot on the top indicates that the measured engine input "start of injection" is practically the same as its reference value, since it is directly derived from static maps. Also the measured "injection duration" is obtained from static maps, but it has a slightly more pronounced deviation from the reference profile (second sub-plot). The engine input "rail pressure" is regulated by means of a fast closed-loop controller, thus peaks appear due to the dynamics characteristics of the system (third sub-plot). Also the input "air mass flow" is regulated by means of a closed-loop controller, which is slower, however, than that for the injection pressure regulation. Peaks due to the dynamics characteristics of the system appear; furthermore, the difference between the signals obtained from the standard measurement with EGR (grey solid line) and the measurement

4.2. Virtual Sensors for λ and NOx

without EGR (black dashed line) is evident: In the second case the normalized air mass flow is much greater than zero, indicating that more air is flowing due to the deactivation of the EGR system (fourth sub-plot). The input "boost pressure" is regulated by means of a yet slower closed-loop controller, and peaks due to the dynamics characteristics of the system appear here as well. A slight difference between measurement with EGR and measurement without EGR is present, because the status of the EGR affects both the air mass flow and the boost pressure control system (fifth sub-plot). Finally, the measured input "engine water temperature" varies slowly and slightly from its reference (sixth sub-plot).

Consider now the parameters for both the λ and the NO_x emission models of Fig. 4.5, which are the same for both measurement with and without EGR. In the case of the λ model (figure on the left), the three parameters have the same order of magnitude. Thus, since the normalized value of the engine input "air mass flow" is greater than those of the other inputs, it has also the greatest influence on the λ estimation. In the case of the NO_x emission model (figure on the right) the parameter of the first sub-plot on the top has the greatest magnitude, but since it relates to the deviation of the start of injection, which is very small, it turns that its influence on the NO_x estimation is moderate. All other parameters are moderate in magnitude, but since the inputs "air mass flow" and "boost pressure" have the greatest normalized values, it turns out that they also have the greatest influence on the NO_x emission estimation.

Figures 4.6 and 4.7 show the λ and NO_x emission profiles, respectively, for both the cases with and without EGR obtained from measurements and from simulations. In addition, the reference profiles are shown. These are obtained simply by linking the corresponding static maps with first-order lag and time delay elements, analogously to Eqs. (4.6) and (4.7), in order to capture the sensors dynamics. In the case of the λ estimation, the reference value correction is mainly due to the normalized air mass flow signal for both measurements with and without EGR. In the case of the NO_x emission estimation, the correction is mainly due to the normalized boost pressure signal for the measurements with EGR and to the normalized air mass flow signal for the measurements without EGR.

The simple idea of Eqs. (4.1)-(4.5) behind the proposed control-oriented models is to "correct" the corresponding reference values by means of a linear combination of the engine inputs. The results described above thus show that it is a good basis to obtain accurate predictions for both the λ value and the NO_x emission, even in the extreme case where the EGR system is completely disabled.

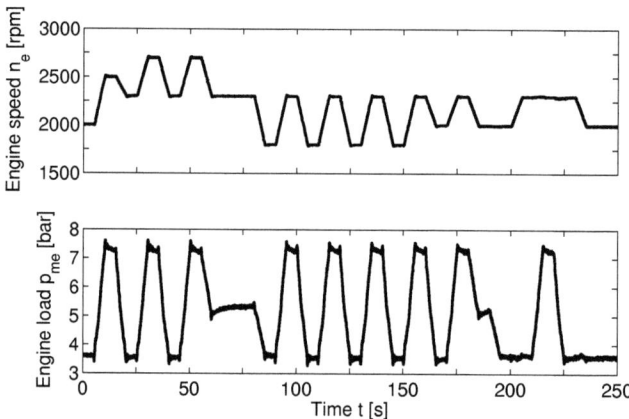

Figure 4.3: Engine speed and load profiles of the example considered.

4.2. Virtual Sensors for λ and NOx

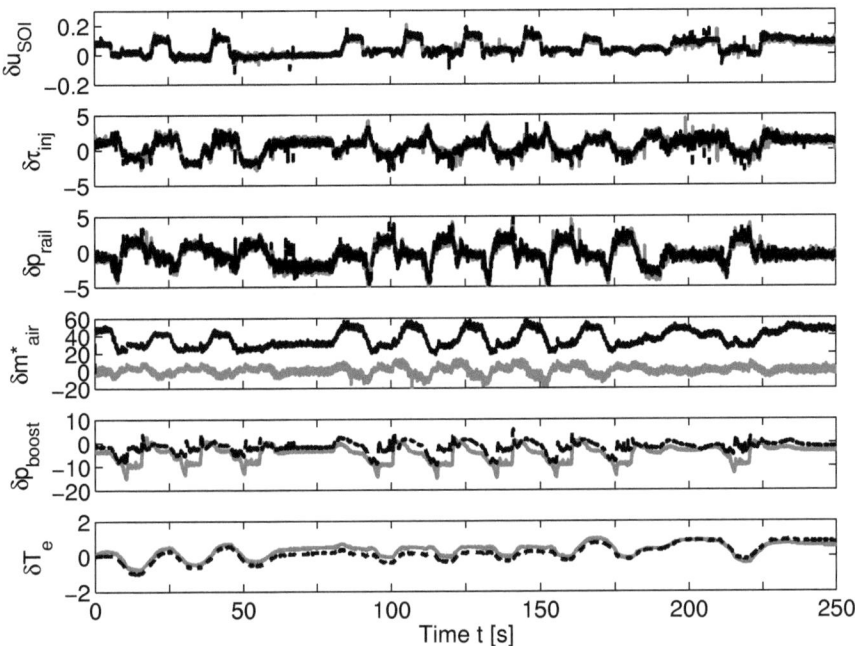

Figure 4.4: Normalized engine inputs (in %) of the example considered, with EGR (grey solid line) and without EGR (black dashed line).

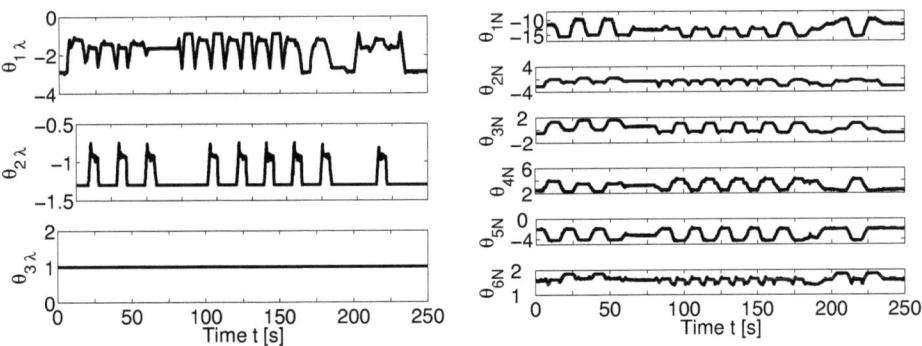

Figure 4.5: Parameters for the λ (left) and the NO_x models (right) of the example considered.

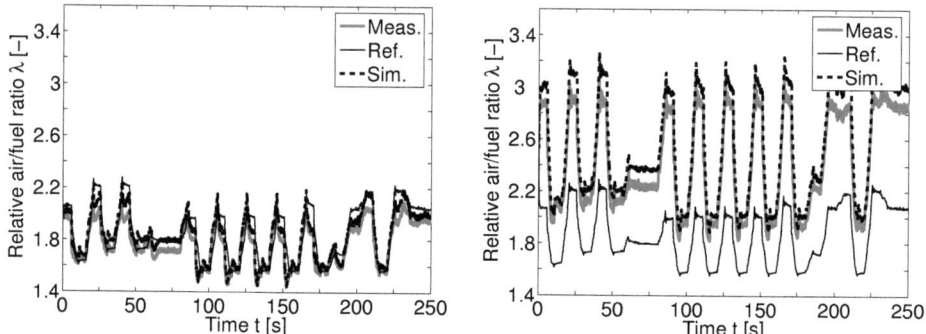

Figure 4.6: Measured and simulated λ values of the example considered, with EGR (left) and without EGR (right).

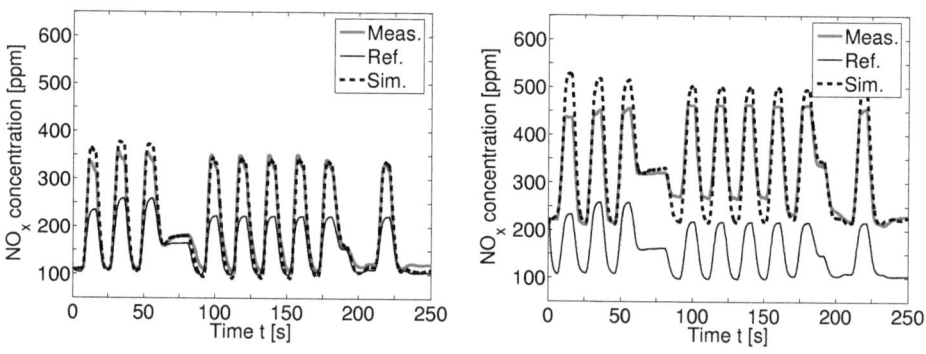

Figure 4.7: Measured and simulated NO_x emission of the example considered, with EGR (left) and without EGR (right).

4.3 Adaptive Virtual NOx Sensor

In contrast to the virtual λ sensor described above, the sensitivities for the virtual NO_x sensor are computed by simulating the complete combustion process model of Chapter 3. Obviously, this model contains unavoidable modeling uncertainties, which consequently affect the complete calculation of the sensitivities. Moreover, whereas for the case of the virtual λ sensor, the linear formulation of Eq. (4.3) is justified by physics, for the virtual NO_x sensor this approach represents only an approximation of the real behavior, which is known to be non-linear. These problems reduce the accuracy of the virtual NO_x sensor.

An efficient solution to improve its accuracy is to apply an adaptive strategy. The idea is to correct the values of some of the model parameters by comparing the model outputs with information obtained from on-line measurements, see Fig. 4.8. The parameters of the virtual NO_x sensor are initialized according to the method described in Section 4.2.2, which leads already to a reasonable parameter set since it has been obtained on the basis of the detailed combustion model, and thus the correction due to the adaptation is expected to be moderate. The virtual NO_x sensor would individually be optimized to each single engine produced, taking also into account the engine variability due to production tolerances, and consequently leading to an engine-specific model.

Various solutions on adaptive model applications have been proposed in the literature. In [42] a sophisticated adaptive NO model for spark-ignition engines is presented. The complete model includes the calculation of the heat release profile, the estimation of the cylinder pressure, the process simulation, and the calculation of the NO formation based on the Zeldovich mechanism. The adaptation is performed using the crankshaft velocity signal, which is used to adjust the parameters for the heat release profile calculation. The model is promising for RT purposes, but it has never been tested on a real ECU. Furthermore, its application to diesel engines requires the development of alternative formulations for the heat release profile calculation that

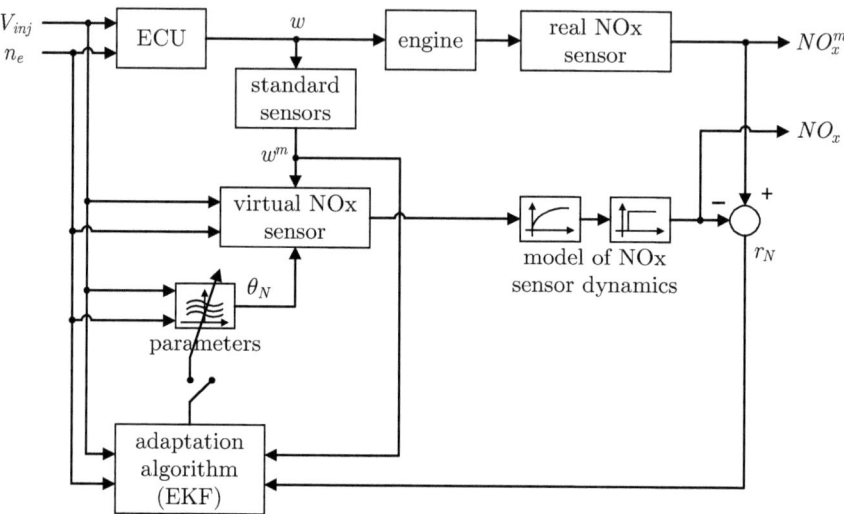

Figure 4.8: The on-line adaptation of the parameter maps of the NO_x model.

are suitable for those engines.

In [77] a control strategy for an LNT system based on an adaptive phenomenological model and the feedback of the signal from a heated exhaust gas oxygen (HEGO) sensor is presented, and results from a simulation model are shown.

4.3.1 Adaptation as a Prerequisite to FDI

Within the framework of this work, the adaptation procedure is thought to be a "prerequisite" for the FDI system, as Fig. 4.9 illustrates. When the engine is new the probability to have a fault is low, consequently the FDI is not necessary yet, and thus the adaptation procedure can be activated: The real NO_x sensor is used as a support to slowly adapt the parameters of the virtual NO_x sensor, adjusting it to the characteristics of the engine. Once the parameters are adjusted the adaptation procedure is deactivated, and at this point the model

4.3. Adaptive Virtual NOx Sensor

matches the real, still fault-free, engine. Subsequently, the FDI system is activated, and the diagnosis is performed on the basis of the adapted model, which is used now as a reference to represent the characteristics of the fault-free engine.

The moment at which the adaptation procedure should be deactivated and the FDI system is to be activated can be considered as a free tuning parameter. For the test-bench experiment results presented in this work, the criterion chosen is the accuracy improvement rate of the virtual NO_x sensor[2]: The adaptation is performed as long as the accuracy of the virtual NO_x sensor is improved, and when the accuracy no longer improves, the adaptation is deactivated. For the operation on a real vehicle this moment should be chosen with great care, since on the one hand too short an adaptation period results in a less accurate NO_x model, but on the other hand, the adaptation period should be interrupted before the deterioration of engine parts becomes a relevant influence. The criterion could be a combination between accuracy improvement rate and the number of kilometers driven by the vehicle.

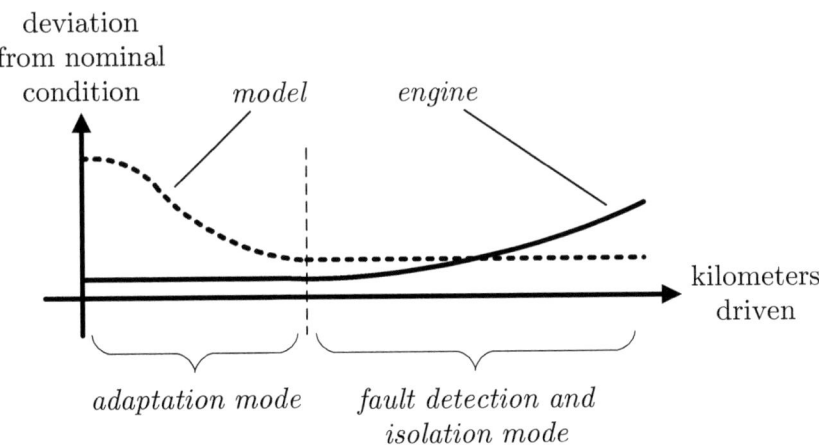

Figure 4.9: The adaptation process as a method to improve the accuracy of the NO_x model and to enhance the FDI performance.

---

[2]Further details are given in Section 6.2.1, Fig. 6.9.

4.3.2 Choice of the Parameters to be Adapted

In practice, the number of parameters that could be efficiently adapted with an identification process is limited by various factors, e.g. modeling uncertainties or lack of excitation. In the case of the present virtual NO_x model the adaptation of all the six parameters could be problematic, thus a choice is made about the best parameters to be adapted.

According to Appendix B.4, the system has to be sufficiently excited to ensure the convergence of the EKF-based identification to the "true" parameters, and the magnitude of this excitation is equivalent to the number of frequencies contained in the signals of the input vector u. The inputs "start of injection" and "injection duration" are almost instantaneous since they are directly calculated from maps, whereas the "cooling water temperature" varies too slowly and can be treated as a constant. Hence, those signals have small frequency content and thus are not suitable for any identification process. The inputs "rail pressure," "air mass flow," and "boost pressure" are regulated by feedback controllers tuned with different time constants. This means that the corresponding signals have sufficient frequency content and thus a certain level of excitation. Moreover, since those three inputs are representative of both the air and the fuel path because they determine the amount of air, recirculated gas, and fuel entering the cylinders, they are of considerable importance.

Therefore, only the three parameters θ_{3N}, θ_{4N}, and θ_{5N} concerning the signals "rail pressure," "air mass flow," and "boost pressure" are adapted.

4.3.3 Adaptation Algorithm

In order to slightly adjust the maps of the parameters to be adapted on-line, the adaptation algorithm is based on the EKF approach described in Appendix B. It is designed according to the guidelines presented in Appendix B.3 to be a slow process.

Several papers on on-line maps adaptation have been published.

4.3. Adaptive Virtual NOx Sensor

For a general introduction about identification algorithms, interested readers are referred to [8]. In [36] a method is proposed for adapting the map of the correction factor for the SOI time estimation based on measurements of injection pressure and quantity of fuel injected. In [114] an algorithm based on the recursive least-square (RLS) technique for the adaptation of the friction torque look-up table based on the measurement of the engine speed is presented. In [79] a comparison between standard RLS-based algorithms and a new proposed one for the adaptation of the ignition angle map of SI engines is presented. First, the cylinder pressure signal obtained with a cylinder pressure sensor is evaluated in order to compute the crank angle location of the 50% energy conversion. Then, the adaptation strategy modifies the map for the point of ignition in order to maintain a constant location of 50% energy conversion of 8° crankshaft angle. That strategy is motivated by thermodynamical analysis as well as experimental results.

The complete algorithm proposed here is composed of the four sequential processes described below. The sequence is repeated every 20 ms, i.e., with the same sampling time as the ECU.

Step 1: Define extended system matrices

First, the state vector is augmented by the parameters to be adapted, which yields the vector $\bar{x}$:

$$\bar{x}(k) = \begin{bmatrix} NO_x(k) \\ \theta_{3N}(k) \\ \theta_{4N}(k) \\ \theta_{5N}(k) \end{bmatrix}. \tag{4.20}$$

This leads to the following extended state-space representation of the extended system:

$$\begin{cases} \bar{x}(k+1) = \bar{A}(k) \cdot \bar{x}(k) + \bar{B}(k) \cdot \bar{u}(k) + \bar{B}_v(k) \cdot \bar{v}(k) \\ y(k) = \bar{C} \cdot \bar{x}(k-n_d) + q(k) \end{cases} \tag{4.21}$$

The corresponding extended system matrices are:

$$\bar{A}(k) = \begin{bmatrix} A_x & A_\theta(k) \\ 0 & 1 \end{bmatrix}, \quad \bar{C} = \begin{bmatrix} C_x & C_\theta \end{bmatrix},$$

$$\bar{B}(k) = \begin{bmatrix} b_N \cdot NO_x^*(k) \cdot \bar{\bar{B}}_\theta(k) \\ 0 \end{bmatrix}, \quad (4.22)$$

where

$$A_x = \frac{\partial f_N}{\partial NO_x} = -a_N \quad (4.23)$$

$$A_\theta(k) = \left[\frac{\partial f_N}{\partial \theta_{3N}}, \frac{\partial f_N}{\partial \theta_{4N}}, \frac{\partial f_N}{\partial \theta_{5N}} \right] = [b_N \cdot NO_x^*(k) \cdot \delta p_{rail}(k), \dots$$
$$\dots b_N \cdot NO_x^*(k) \cdot \delta \dot{m}_{air}(k), \dots$$
$$\dots b_N \cdot NO_x^*(k) \cdot \delta p_{boost}(k)] \quad (4.24)$$

$$C_x = \frac{\partial g_N}{\partial NO_x} = 1 \quad (4.25)$$

$$C_\theta = \left[\frac{\partial g_N}{\partial \theta_{3N}}, \frac{\partial g_N}{\partial \theta_{4N}}, \frac{\partial g_N}{\partial \theta_{5N}} \right] = [0,\ 0,\ 0] \quad (4.26)$$

and

$$\bar{\bar{B}}_\theta(k) = \begin{bmatrix} 1 & \theta_{1N}(k) & \theta_{2N}(k) & 0 & 0 & 0 & \theta_{6N}(k) \end{bmatrix}, \quad (4.27)$$

i.e., those parameters that are adapted and thus appear in the extended state vector $\bar{x}$, are eliminated from the extended system input matrix $\bar{B}$. Since the matrix $\bar{B}_v$ is chosen to be

$$\bar{B}_v(k) = \begin{bmatrix} B_v(k) \\ 0 \end{bmatrix}, \quad (4.28)$$

the input noise in not acting on the parameters to be adapted.

4.3. Adaptive Virtual NOx Sensor

Hence expanding Eq. (4.21) one can obtain:

$$\begin{bmatrix} NO_x(k+1) \\ \theta_{3N}(k+1) \\ \theta_{4N}(k+1) \\ \theta_{5N}(k+1) \end{bmatrix} = \begin{bmatrix} -a_N & b_N NO_x^*(k)\delta p_{rail}(k) & b_N NO_x^*(k)\delta \dot{m}_{air}(k) & \cdots \\ 0 & 1 & 0 & \cdots \\ 0 & 0 & 1 & \cdots \\ 0 & 0 & 0 & \cdots \end{bmatrix}$$

$$\begin{matrix} \cdots & b_N NO_x^*(k)\delta p_{boost}(k) \\ \cdots & 0 \\ \cdots & 0 \\ \cdots & 1 \end{matrix} \begin{bmatrix} NO_x(k) \\ \theta_{3N}(k) \\ \theta_{4N}(k) \\ \theta_{5N}(k) \end{bmatrix} +$$

$$+ \begin{bmatrix} b_N NO_x^*(k) & b_N NO_x^*(k)\theta_{1N}(k) & b_N NO_x^*(k)\theta_{2N}(k) & 0 & 0 & 0 & \cdots \\ 0 & 0 & 0 & 0 & 0 & 0 & \cdots \\ 0 & 0 & 0 & 0 & 0 & 0 & \cdots \\ 0 & 0 & 0 & 0 & 0 & 0 & \cdots \end{bmatrix}$$

$$\begin{matrix} \cdots & b_N NO_x^*(k)\theta_{6N}(k) \\ \cdots & 0 \\ \cdots & 0 \\ \cdots & 0 \end{matrix} \begin{bmatrix} 1 \\ \delta u_{SOI}(k) \\ \delta \tau_{inj}(k) \\ \delta p_{rail}(k) \\ \delta \dot{m}_{air}(k) \\ \delta p_{boost}(k) \\ \delta T_e(k) \end{bmatrix} + \bar{B}_v(k)\bar{v}(k). \quad (4.29)$$

Step 2: Read maps of parameters and covariance matrix elements and define extended state vector and covariance matrix

Successively, the parameters to be adapted and all elements of the covariance matrix are read from the corresponding maps as a function of the actual operating conditions in order to build the actual estimate of the state vector $\hat{x}$ and the covariance matrix Σ, as shown in Fig. 4.10.

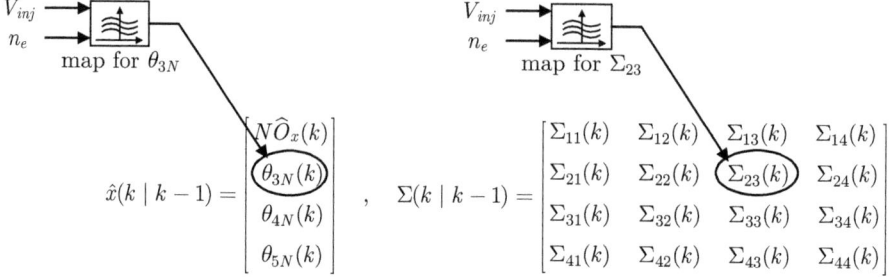

Figure 4.10: The determination of the parameter θ_{3N} and of the element Σ_{23} is shown exemplarily.

Step 3: Update state vector and covariance matrix

The core of the adaptation procedure is represented by the update of the state vector and the covariance matrix, which is performed by means of an EKF. The EKF algorithm described in Appendix B is specified by the computation of the estimation residual on the basis of the measurements NO_x^m from the NO_x sensor and of the estimate of the state vector $\hat{x}$ (Eq. (4.30)), the "measurement update" or "correction step" (Eq. (4.31)), and the "time update" or "extrapolation step" (Eq. (4.32)):

$$r_N(k) = NO_x^m(k) - \bar{C} \cdot \hat{x}(k|k-1) \qquad (4.30)$$

$$\begin{aligned}
Q(k) &= \bar{C} \cdot \Sigma(k|k-1) \cdot \bar{C}^T + R_q \\
L(k) &= \Sigma(k|k-1) \cdot \bar{C}^T \cdot Q^{-1}(k) \\
\hat{x}(k|k) &= \hat{x}(k|k-1) + L(k) \cdot r_N(k) \\
\Sigma(k|k) &= \Sigma(k|k-1) - L(k) \cdot \bar{C} \cdot \Sigma(k|k-1)
\end{aligned} \qquad (4.31)$$

$$\begin{aligned}
\hat{x}(k+1|k) &= \bar{A}(k) \cdot \hat{x}(k|k) + \bar{B}(k) \cdot \bar{u}(k) \\
\Sigma(k+1|k) &= \bar{A}(k) \cdot \Sigma(k|k) \cdot \bar{A}^T(k) + \bar{B}_v(k) \cdot R_v \cdot \bar{B}_v^T(k).
\end{aligned} \qquad (4.32)$$

4.3. Adaptive Virtual NOx Sensor

Step 4: Write updated parameters and covariance matrix elements back into the maps

At the end of the adaptation procedure the new updated parameters and all elements of the covariance matrix are saved in the corresponding maps as a function of the actual operating conditions, as shown in Fig. 4.11.

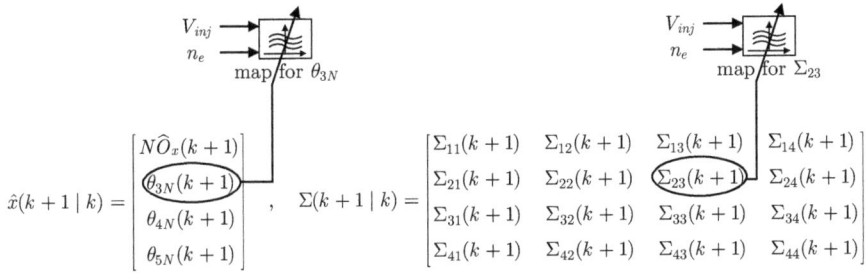

Figure 4.11: The update of the maps for the parameter θ_{3N} and for the element Σ_{23} is shown exemplarily.

4.3.4 Example

The proposed algorithm is tested on the basis of dynamic measurements conducted using the experimental facilities mentioned above. Figure 4.12 shows the measured and the simulated NO_x profiles for load variations between 2 and 6 bar BMEP at a constant engine speed of 2000 rpm. At the beginning of the experiment, the parameters for the virtual NO_x sensor are those obtained with the approach described in Section 4.2.2. First, the adaptation loop is disabled. After about 100 s, the adaptation loop is enabled and, finally, after another 100 s, it is disabled. The positive effect of the adaptation is clear: During the first 100 s, when the parameters are not adapted yet, the error between real and virtual NO_x sensor (completed with lag element and delay) is greater than during the last 100 s, when the parameters are adapted.

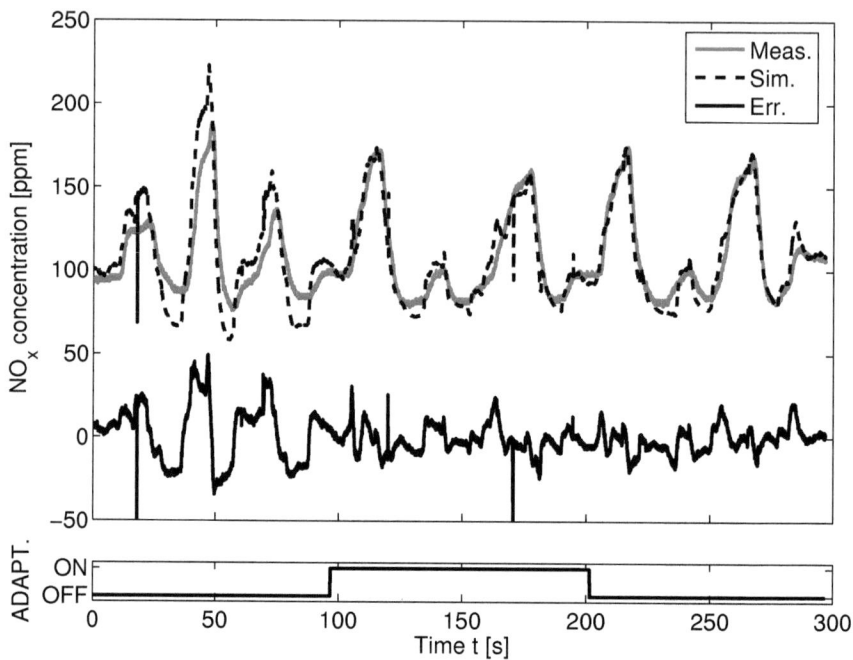

Figure 4.12: Profiles of the NO_x emission showing load variations at constant engine speed, with adaptation procedure enabled and disabled.

Another experiment is performed in accordance with the engine speed and load profile of Fig. 4.13. Again, at the beginning of the experiment, the parameters for the virtual sensor are those obtained with the approach described in Section 4.2.2. Figure 4.14 shows the corresponding parameters before (thin solid line) and after (thick dashed line) the adaptation. Figure 4.15 shows the NO_x profiles measured with the real sensor and simulated with the virtual sensor (completed with lag element and delay), before and after the adaptation process. Clearly, the effect of the adaptation on the NO_x prediction quality is positive.

4.3. Adaptive Virtual NOx Sensor

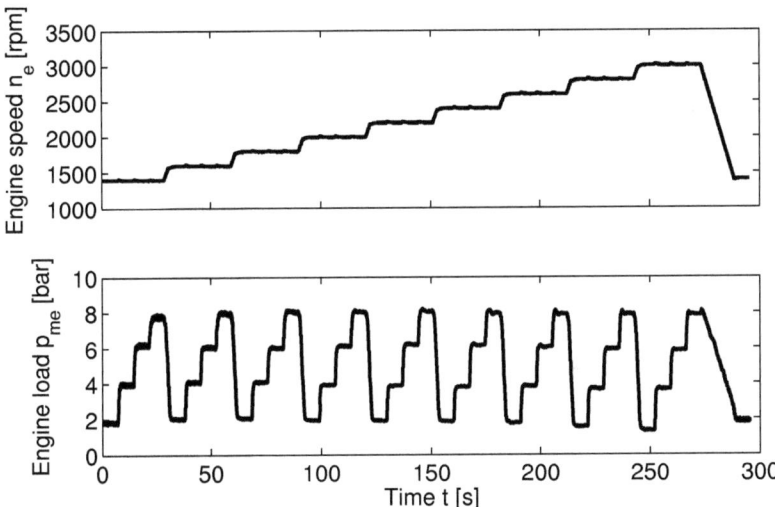

Figure 4.13: Engine speed and load profiles of the example considered.

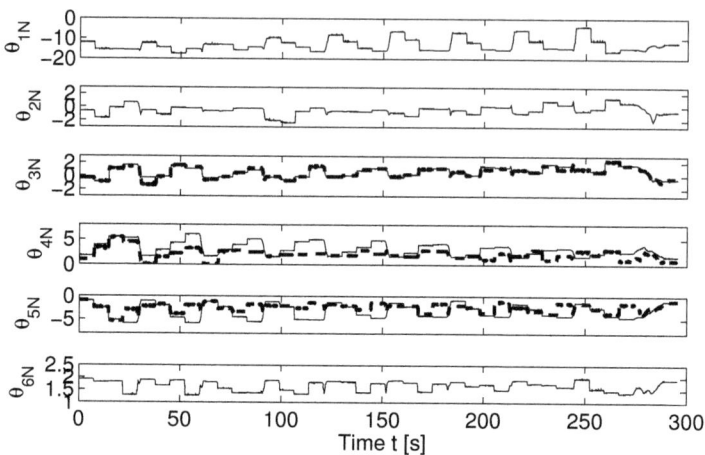

Figure 4.14: The six NO_x model parameters before (thin solid line) and after (thick dashed line) the adaptation procedure.

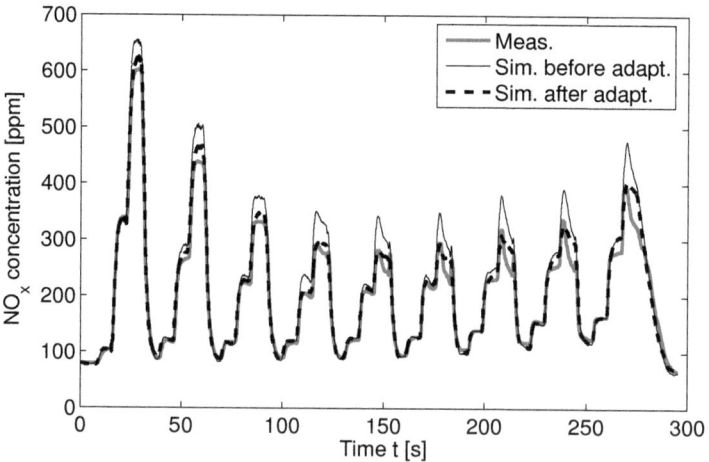

Figure 4.15: Profiles of the NO_x emission before and after the adaptation procedure for the example considered.

4.4 Application of the Adaptive Virtual NOx Sensor to a Series-Production Engine

This section is intended to demonstrate the application of the complete proposed methodology to obtain an accurate real-time NO_x model to a different engine than the one specified in Table 2.1. Results from a realistic situation typical of the automotive industry are presented. The methodology includes the calibration of the combustion model with static measurements, the real-time NO_x model parameter calculation, and the adaptation of the NO_x model on the basis of dynamic measurements, according to Fig. 1.5.

The development and test conditions described in the previous sections are ideal, i.e., with free and unlimited access to the test-bench, optimal instrumentation, and with the same engine used for the calibration of the combustion model (static measurement) as for the experimental validation of the virtual NO_x sensor (dynamic measurement). In the reality of the automotive industry this is rarely the case. Test-bench time and development costs are limiting factors. Typically, one single engine of a given class is optimized and tuned on a test-bench and, successively, each engine of this class that is produced is adapted to the different vehicle categories with their slightly different characteristics. Furthermore, each individual engine shows variability phenomena due to production tolerances that slightly change its characteristics in an unpredictable manner. For these reasons it is often not possible to develop a unique model valid for all vehicles equipped with the same engine.

The proposed adaptive virtual NO_x sensor could solve this problem. Starting from the general real-time model developed in Section 4.2, the algorithm adapts the parameters of each engine of the same class individually. Thus the initial parameters are the same for all engines of the same class, but during the adaptation process they are adjusted individually.

The task is the application of the algorithms presented until now to a completely different diesel engine class than the one defined in Table 2.1 using only information provided by the engine manufacturer. Details about this alternative engine class are given in Table 4.2. Data concerning the engine geometry, the injector geometry, and the valve timing, as well as data from static test-bench measurements and dynamic measurements obtained from a real vehicle were available. All the measurement data were recorded by the manufacturer, and the engine unit of the vehicle used for the dynamic measurement was different from those used for the static measurement.

Table 4.2: Technical data of the engine used to test the proposed methodology on the basis of measurement data provided by the engine manufacturer.

type	direct-injection, common-rail
features	EGR, VNT and IPSO
architecture	6 cylinders, 24 valves
bore x stroke	83 mm x 92 mm
compression ratio	18:1
maximum torque	540 Nm at 1600-2400 rpm
maximum power	173 kW at 3600 rpm
emissions level	EURO IV

The engine, injector, and valve data, as well as the data from static measurements as shown in Fig. 4.16, are used to calibrate the detailed combustion model. The results are shown in Figs. 4.18 and 4.19. The parameters for the virtual NO_x sensor are then initialized according to the strategy described in Section 4.2.2. Finally, the adaptation process is tested by means of the dynamic measurement shown in the traces of Fig. 4.17. The comparison between the parameters before (thin solid line) and after (thick dashed line) the adaptation is shown in Fig. 4.20. Figure 4.21 shows the NO_x profiles measured with the real sensor and simulated with the virtual sensor (completed with lag element and delay), before and after the adaptation process. Clearly, the effect of

4.4. Application to a Series-Production Engine 81

the adaptation on the NO_x prediction quality is positive.

Figure 4.21 also shows that the simulation before the adaptation procedure is affected by a noticeable error, compared to the example conducted with the test-bench engine shown in Fig. 4.15. This error correlates well with the accuracy of the combustion model used to calculate the parameters, as confirmed by the comparison between Figs. 3.15 right and 4.19 right. This fact points out the advantages of the adaptation procedure proposed here: Using this adaptation algorithm an accurate NO_x model can be successfully obtained, even if the initial parameter set calculated with the combustion model is uncertain.

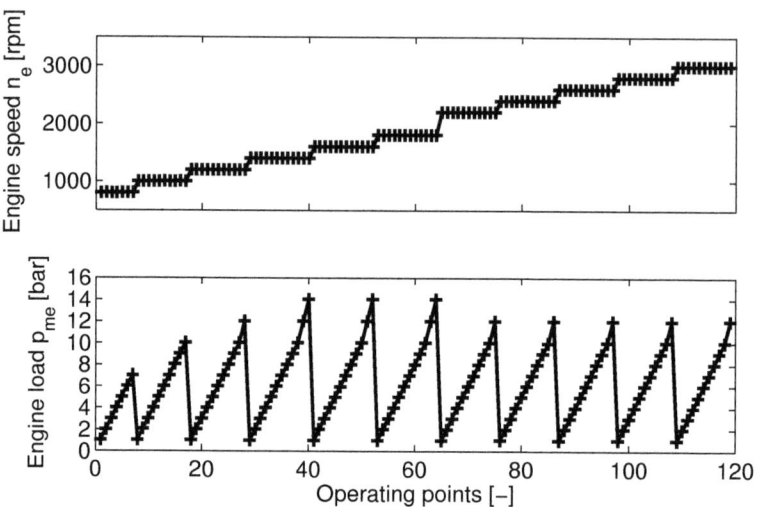

Figure 4.16: Static operating points received from the engine manufacturer, used to calibrate the combustion model.

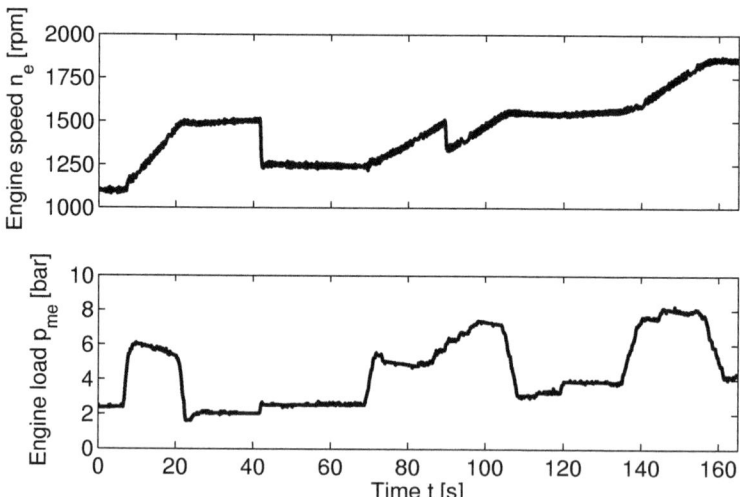

Figure 4.17: Engine speed and load profiles received from the engine manufacturer.

4.4. Application to a Series-Production Engine

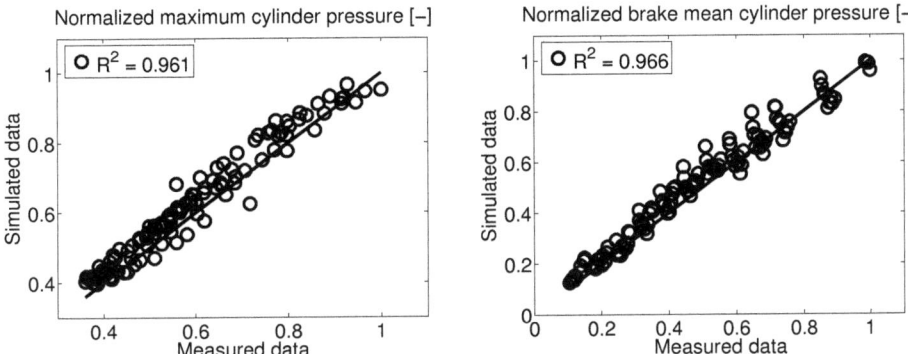

Figure 4.18: Comparison between measured and simulated normalized maximum cylinder pressure (left) and normalized indicated mean effective cylinder pressure (right).

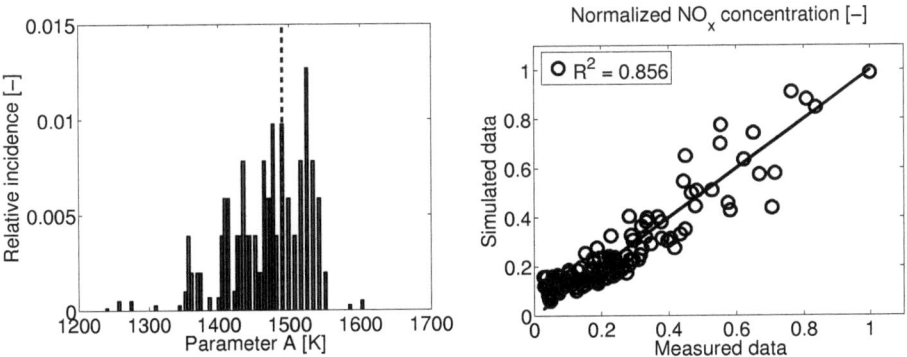

Figure 4.19: Determination of the A-parameter for the engine of Table 4.2 (left), and comparison between measured and simulated normalized NO_x concentration levels (right).

84   *Chapter 4.   Control-Oriented Emission Models*

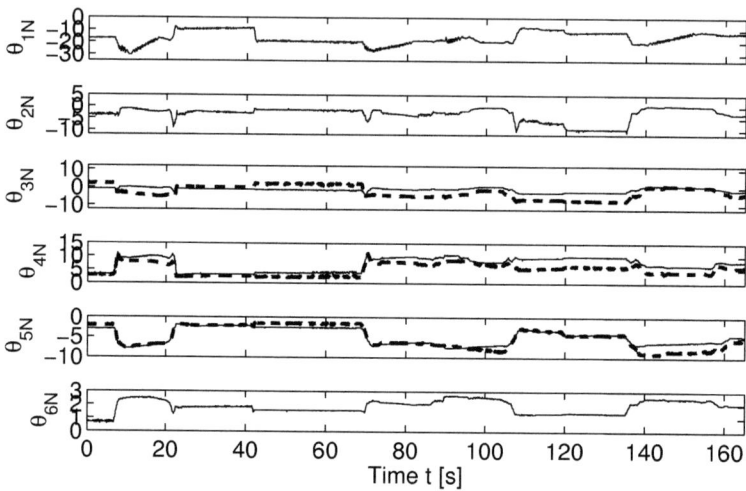

Figure 4.20: The six NO_x model parameters before (thin solid line) and after (thick dashed line) the adaptation procedure.

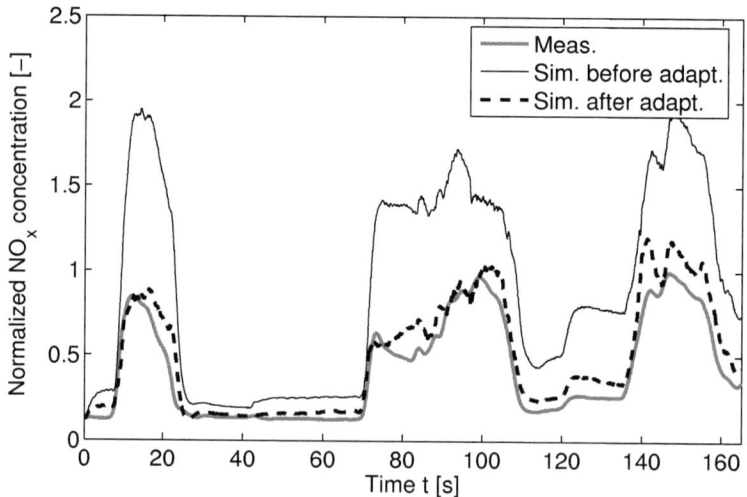

Figure 4.21: Profiles of the NO_x emission before and after the adaptation procedure for the example considered.

Chapter 5

Model-Based Fault Detection and Isolation of a DI Diesel Engine

The system developed here is based on the proposed λ and NO_x control-oriented models, as well as on the corresponding sensor signals. Both models are first used in a "forward" manner to determine whether a fault is present or not, and successively they are used in a "backward" manner to quantify and localize the fault.

5.1 Contribution of the Proposed Automotive FDI System

To contextualize and clarify the contribution of the proposed FDI system, the relevant work done on automotive FDI for both SI and diesel engines is briefly introduced. Successively, the salient characteristics of the proposed work are presented, discussed, and linked to the previous work on automotive FDI.

5.1.1 Previous Automotive FDI Systems

The research on FDI systems for automotive engines has been driven mainly by legislation, with the introduction of OBD regulations, and has been made possible with progess in technology, i.e., with the development of increasingly powerful microprocessors.

The development of the OBD regulations can be briefly summed up as follows: First, the OBD-1 standard was introduced in 1987 by the California Air Resource Board (CARB) for all vehicles sold in California, then in 1996 the OBD-2 standard, which is an extension of OBD-1, became mandatory for all vehicles sold in the United States. The European on-board diagnostics standard, called E-OBD, is a variant of the OBD-2, and was introduced in 2001 for SI engines and in 2003 for diesel engines.

The first work on automotive FDI was mainly oriented to SI engines, whereas diesel engines are nowadays considered with increasing interest for research and development of FDI systems. In most of the work presented the FDI systems are based on models of the engine, and the fault detection task is performed by means of parity equations and threshold checking. However, approaches based on signal analysis, which have been developed recently, were investigated also.

At the beginning the models were based on black-box approaches, for instance simple empirical polynomial functions whose parameters were determined using identification methods. Successively the models became physics-based or, at least, contained both physical representations of some processes and black-box formulations. Also black-box approaches based on NN have been explored especially in the last years, since NN are a powerful tool to map the non-linear characteristics of engines.

SI Engines Academic Research

One of the first contributions to automotive FDI of SI engines is the one proposed in [69], where a model-based FDI system for the detection of failures in the throttle position and in the manifold pressure sensors

5.1. Contribution

is presented. A linear empirical engine model is used in this case.

Other early contributions are those proposed in [44] and [45], where an algorithm is presented that is based on the structured parity equation methodology for the on-line detection and diagnosis of faults in the fuel injectors, EGR valve, throttle position sensor, manifold pressure sensor, engine speed sensor, and exhaust oxygen sensor. The models of the engine are represented by polynomials whose parameters are found by identification.

Alternative approaches have been investigated also: In [52] a model-based diagnosis system based on a sliding-mode methodology for non-linear estimation of the engine inputs and outputs is proposed. Residuals are generated by means of a model based on principles described in [4].

In [75] an interesting method is presented for diagnosing air mass flow sensor fault, throttle position sensor fault, intake manifold leakage, injector fault, misfire, and EGR valve fault based on fuzzy inference. However, that method relies on exhaust gas measurements (CO, NO_x, HC) that are available only off-line.

Further relevant contributions are those proposed in [59], [60], and [109], where a diagnosis method is presented based on a structure of hypothesis testing for the detection of faults in the air mass sensor, throttle angle sensor, and manifold pressure sensor. In those cases, the models of the engine were developed following both physical and black-box approaches. That work is completed with the results shown in [61], where two methods for diagnosing leakages in the air path of automotive engines based on hypothesis testing are investigated: The first relies on the comparison between measured and estimated air flow, and the second on an estimation of the leakage area. In that case also, the modeling is based on both physical and black-box approaches.

Diesel Engines Academic Research

An interesting application of the NN technique to the automotive FDI of diesel engines is shown in [57], where a method for detecting faults of the fuel injection system based on the measured cylinder pressure signal is presented. First, the difference pressure, i.e., the pressure rise due to combustion, is derived using the measured cylinder pressure signal. Then, features are extracted from the difference pressure. Successively, the extracted features are used as inputs for an NN in order to estimate the fuel mass injected and the injection angle. Finally, detection is performed by comparing the estimated variables with reference values, and implementing thresholds.

Obviously, the approaches to develop FDI systems for diesel engines could also be inspired by those proposed for SI engines. That is the case of [62], where a model-based diagnosis system is proposed for the air path of diesel engines able to diagnose air mass flow sensor fault, intake manifold pressure sensor fault, air leakage, and EGR valve fault. The fault isolation is performed with structured hypothesis tests. The model used in that case is based on principles described in [4].

Another contribution based on NN is presented in [53], [74], and [121], where a model-based fault detection and diagnosis system has been proposed. Various fault modes have been considered: Removed tube of the crank case vent, leakage between intercooler and engine, restriction between intercooler and engine, swirl flap actuator fault, and EGR valve fault. The faults are detected by analyzing residuals obtained from parity equations. Where a physics-based approach was not possible, reference models are developed in a hybrid fashion, i.e., linking physics-based equations with black-box formulations based on the local linear neural networks presented in [47]. However, the presented approach has not been tested on current ECUs.

As mentioned before, novel fault detection methods based on signal analysis have been recently developed: In [64] a method for the diagnosis of the injection system through common-rail pressure is presented. The injection pressure signal is evaluated by means of a short-time Fourier transformation, and the detection of a total lack of injection of one cylinder is performed by limit-checking the evaluated injection pressure signal.

In [98] a method based on singular spectrum analysis to detect fault of the charge air cooler is proposed. This tool is commonly used for various applications to smooth signals and remove noise. For the

5.1. Contribution 89

specific problem described here this is a very useful feature, since the changes in the signals when the failure is present are subtle and thus difficult to identify. Using singular spectrum analysis signal detection, singular values from fault data are compared against healthy data. The discrepancy is then evaluated by means of a statistical significance testing in order to determine whether a fault is present or not.

5.1.2 Proposed Automotive FDI System

From this brief literature overview on automotive FDI systems, it turns that the research on FDI of diesel engines is currently extremely vital, and it is one of the aims of this work to contribute to its vitality.

The main characteristics of the proposed FDI system is the use of measurement signals of engine-out values, obtained from solid-state on-line λ and NO_x sensors, see Fig. 5.1. This is a novelty in the field of automotive FDI, since all work done until now has been based on ECU measurement or cylinder pressure signals. The approach proposed follows a model-based strategy, where the fault detection is performed by means of parity equations and threshold checking. The on-line models proposed are new as well, since they are a linear formulation of the two engine-out variables λ and NO_x. Their parameters (sensitivities) are calculated by means of a more detailed physics-based model of the combustion process developed within the scope of this work.

The three faults considered concern the engine inputs "injected fuel quantity," "air mass flow," and "boost pressure." The reason for this choice is that air mass flow and boost pressure are representative values for the air path and injected fuel quantity for the fuel path. They are the most important quantities for the definition of the amount of air, of EGR, and of fuel entering the cylinders. Moreover, since the air mass flow is controlled in a closed loop, as shown in Fig. 5.1, it is not relevant for the purpose of estimating its exact level, whether the corresponding fault mode is caused by a fault of the air mass flow sensor or a fault of the EGR valve. Analogously, the boost pressure is also controlled in a closed loop, as shown in Fig. 5.1, and thus it is not relevant for the purpose of estimating its exact level, whether

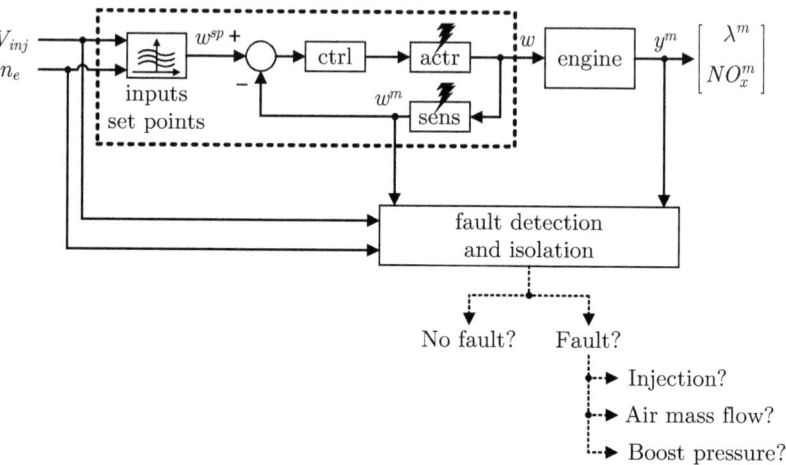

Figure 5.1: The proposed system to detect and isolate faults of the air and fuel paths embedded in a standard arbitrary diesel engine control loop, where "ctrl" = controller, "actr" = actuator, and "sens" = sensor. The flashes indicate the components that could cause faults.

the corresponding fault mode has been caused by a fault of the boost pressure sensor or a fault of the VNT actuator. The quantity of fuel injected is proportional to the gas pedal position, and of course it is influenced by the status of the injectors, which are known to be engine parts that easily deteriorate (e.g. fouling).

A further novelty of the proposed FDI system is thus the possibility to simultaneously perform the diagnosis of both the air and fuel paths, whereas previous approaches focused on the diagnosis of the air path or of the fuel path.

5.2 Fault Detection

For the fault detection, the measured λ and NO_x values y^m are compared with the corresponding calculated values y of the control-

5.3. FDI Strategy A

oriented models and normalized:

$$r_{FD}(k) = \frac{y(k) - y^m(k)}{y(k)} \quad (5.1)$$

The raw normalized residuals of Eq. (5.1) obtained are then filtered by means of first-order lag elements. The condition for the fault detection is straightforward: As soon as one of the filtered residuals exceeds the threshold imposed by the designer, a fault is detected. Figure 5.2 shows exemplarily both λ and NO_x residuals for a fault of the air mass flow.

The normalization of the residuals is introduced to improve the fault detection capabilities, and it is equivalent to implementing adaptive thresholds, see [7] and [10]. In this case the solution adopted to implement adaptive thresholds is quite simple and intuitive, since due to normalization the residuals are proportional to the magnitude of the measured sensors signals. While the thresholds are tuning parameters for the FDI system, it still makes sense to choose values similar to those defined in the accuracy specifications of the sensors, as listed in Tables 2.4 and 2.5.

The performance of the fault detection algorithm could be further improved by introducing a second layer of filtering and an additional round of threshold testing, as proposed in [45], in order to reduce false alarms. In this work, however, this last solution has not been implemented.

5.3 FDI Strategy A: One EKF

Once a fault is detected, it has to be quantified and localized. In the "FDI strategy A" the λ and NO_x measurements are used to reconstruct the engine inputs suspected to be faulty by inverting the control-oriented models. This is done by means of one single EKF, which provides a simultaneous estimate of the three engine inputs assumed to be faulty, as shown in Fig. 5.3. Ideally, the EKF estimate of the two non-faulty engine inputs should be zero, whereas the one

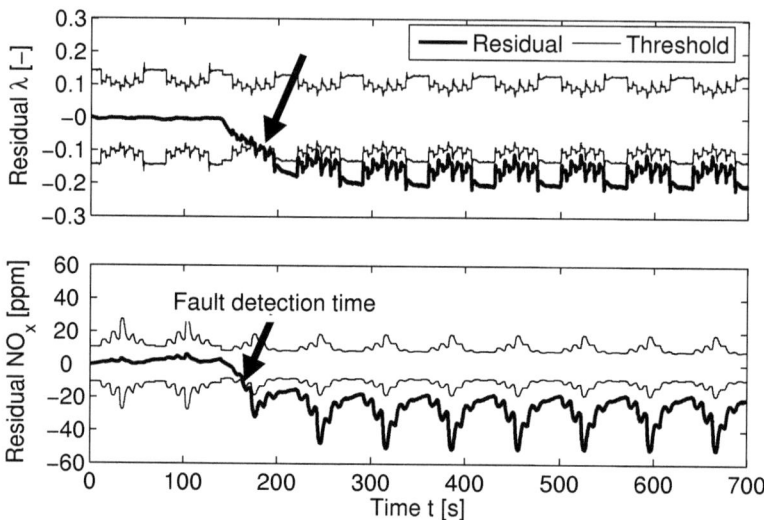

Figure 5.2: Normalized and filtered residuals of λ and NO_x for a fault of the air mass flow of -10% introduced at $140\ s$, and corresponding adaptive thresholds.

of the faulty one should clearly not be equal to zero. Thus the fault estimation and classification processes are executed simultaneously.

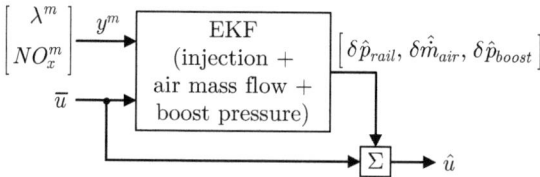

Figure 5.3: The proposed FDI strategy A, based on one single EKF for the simultaneous estimation of the engine inputs "injected fuel quantity," "air mass flow," and "boost pressure."

5.3.1 Fault Estimation and Classification

The state vector is augmented by the inputs to be identified. Accordingly, those inputs that are identified are eliminated from the input vector. Hence the vectors $\bar{u}$ and $\bar{x}$, which describe the modified input and state vectors, are defined by:

$$\bar{u}(k) = \begin{bmatrix} 1 \\ 1 \\ \delta u_{SOI}(k) \\ \delta T_{inj}(k) \\ 0 \\ 0 \\ 0 \\ \delta T_e(k) \end{bmatrix}, \quad \bar{x}(k) = \begin{bmatrix} \lambda(k) \\ NO_x(k) \\ \delta p_{rail}(k) \\ \delta \dot{m}_{air}(k) \\ \delta p_{boost}(k) \end{bmatrix}. \quad (5.2)$$

This leads to the following extended state-space representation of the system:

$$\begin{cases} \bar{x}(k+1) = \bar{A}(k) \cdot \bar{x}(k) + \bar{B}(k) \cdot \bar{u}(k) + \bar{B}_v(k) \cdot \bar{v}(k) \\ y(k) = \bar{C} \cdot \bar{x}(k - n_d) + q(k) \end{cases} \quad (5.3)$$

The corresponding extended system matrices are:

$$\bar{A}(k) = \begin{bmatrix} A_x & A_u(k) \\ 0 & 1 \end{bmatrix}, \quad \bar{B}(k) = \begin{bmatrix} B(k) \\ 0 \end{bmatrix}, \quad \bar{C} = \begin{bmatrix} C_x & C_u \end{bmatrix}, \quad (5.4)$$

where

$$A_x = \begin{bmatrix} \frac{\partial f_\lambda}{\partial \lambda} & \frac{\partial f_N}{\partial \lambda} \\ \frac{\partial f_\lambda}{\partial NO_x} & \frac{\partial f_N}{\partial NO_x} \end{bmatrix} = \begin{bmatrix} -a_\lambda & 0 \\ 0 & -a_N \end{bmatrix}, \quad (5.5)$$

$$A_u(k) = \begin{bmatrix} \frac{\partial f_\lambda}{\partial p_{rail}} & \frac{\partial f_\lambda}{\partial \dot{m}_{air}} & \frac{\partial f_\lambda}{\partial p_{boost}} \\ \frac{\partial f_N}{\partial p_{rail}} & \frac{\partial f_N}{\partial \dot{m}_{air}} & \frac{\partial f_N}{\partial p_{boost}} \end{bmatrix} =$$

$$= \tilde{B}^*(k) \cdot \begin{bmatrix} \theta_{2\lambda}(k) & \theta_{3\lambda}(k) & 0 \\ \theta_{3N}(k) & \theta_{4N}(k) & \theta_{5N}(k) \end{bmatrix}, \tag{5.6}$$

$$C_x = \begin{bmatrix} \frac{\partial g_\lambda}{\partial \lambda} & \frac{\partial g_\lambda}{\partial NO_x} \\ \frac{\partial g_N}{\partial \lambda} & \frac{\partial g_N}{\partial NO_x} \end{bmatrix} = \begin{bmatrix} 1 & 0 \\ 0 & 1 \end{bmatrix}, \tag{5.7}$$

$$C_u = \begin{bmatrix} \frac{\partial g_\lambda}{\partial p_{rail}} & \frac{\partial g_\lambda}{\partial \dot{m}_{air}} & \frac{\partial g_\lambda}{\partial p_{boost}} \\ \frac{\partial g_N}{\partial p_{rail}} & \frac{\partial g_N}{\partial \dot{m}_{air}} & \frac{\partial g_N}{\partial p_{boost}} \end{bmatrix} = \begin{bmatrix} 0 & 0 & 0 \\ 0 & 0 & 0 \end{bmatrix}. \tag{5.8}$$

The matrix $\tilde{B}^*$ has been defined previously in Eq. (4.11). Hence expanding Eq. (5.3) one can obtain:

$$\begin{bmatrix} \lambda(k+1) \\ NO_x(k+1) \\ \delta p_{rail}(k+1) \\ \delta \dot{m}_{air}(k+1) \\ \delta p_{boost}(k+1) \end{bmatrix} = \begin{bmatrix} -a_\lambda & 0 & b_\lambda \lambda^*(k)\theta_{2\lambda}(k) & \cdots \\ 0 & -a_N & b_N NO_x^*(k)\theta_{3N}(k) & \cdots \\ 0 & 0 & 1 & \cdots \\ 0 & 0 & 0 & \cdots \\ 0 & 0 & 0 & \cdots \end{bmatrix}$$

$$\begin{bmatrix} \cdots & b_\lambda \lambda \theta_{3\lambda}(k) & 0 \\ \cdots & b_N NO_x^*(k)\theta_{4N}(k) & b_N NO_x^*(k)\theta_{5N}(k) \\ \cdots & 0 & 0 \\ \cdots & 1 & 0 \\ \cdots & 0 & 1 \end{bmatrix} \begin{bmatrix} \lambda(k) \\ NO_x(k) \\ \delta p_{rail}(k) \\ \delta \dot{m}_{air}(k) \\ \delta p_{boost}(k) \end{bmatrix} +$$

$$+ \begin{bmatrix} b_\lambda \lambda^*(k) & 0 & 0 & b_\lambda \lambda^*(k)\theta_{1\lambda}(k) & \cdots \\ 0 & b_N NO_x^*(k) & b_N NO_x^*(k)\theta_{1N}(k) & b_N NO_x^*(k)\theta_{2N}(k) & \cdots \\ 0 & 0 & 0 & 0 & \cdots \\ 0 & 0 & 0 & 0 & \cdots \\ 0 & 0 & 0 & 0 & \cdots \end{bmatrix}$$

5.3. FDI Strategy A

$$\begin{bmatrix} \cdots & b_\lambda \lambda^*(k)\theta_{2\lambda}(k) & b_\lambda \lambda^*(k)\theta_{3\lambda}(k) & 0 & \cdots \\ \cdots & b_N NO_x^*(k)\theta_{3N}(k) & b_N NO_x^*(k)\theta_{4N}(k) & b_N NO_x^*(k)\theta_{5N}(k) & \cdots \\ \cdots & 0 & 0 & 0 & \cdots \\ \cdots & 0 & 0 & 0 & \cdots \\ \cdots & 0 & 0 & 0 & \cdots \end{bmatrix}$$

$$\left.\begin{matrix} \cdots & 0 & \\ \cdots & b_N NO_x^*(k)\theta_{6N}(k) & \\ \cdots & 0 & \\ \cdots & 0 & \\ \cdots & 0 & \end{matrix}\right] \begin{bmatrix} 1 \\ 1 \\ \delta u_{SOI}(k) \\ \delta \tau_{inj}(k) \\ 0 \\ 0 \\ 0 \\ \delta T_e(k) \end{bmatrix} + \bar{B}_v(k)\bar{v}(k). \quad (5.9)$$

The estimation of the engine inputs is based on the same principle as the adaptation algorithm. Hence also in this case the EKF algorithm according to Appendix B is applied, performing the computation of the estimation residual on the basis of the measurements y^m from the λ and NO_x sensors and of the estimate of the state vector $\hat{x}$ (Eq. (5.10)), the "measurement update" or "correction step" (Eq. (5.11)), and the "time update" or "extrapolation step" (Eq. (5.12)):

$$r(k) = y^m(k) - \bar{C} \cdot \hat{x}(k|k-1) \quad (5.10)$$

$$\begin{aligned} Q(k) &= \bar{C} \cdot \Sigma(k|k-1) \cdot \bar{C}^T + R_q \\ L(k) &= \Sigma(k|k-1) \cdot \bar{C}^T \cdot Q^{-1}(k) \\ \hat{x}(k|k) &= \hat{x}(k|k-1) + L(k) \cdot r(k) \\ \Sigma(k|k) &= \Sigma(k|k-1) - L(k) \cdot \bar{C} \cdot \Sigma(k|k-1) \end{aligned} \quad (5.11)$$

$$\begin{aligned} \hat{x}(k+1|k) &= \bar{A}(k) \cdot \hat{x}(k|k) + \bar{B}(k) \cdot \bar{u}(k) \\ \Sigma(k+1|k) &= \bar{A}(k) \cdot \Sigma(k|k) \cdot \bar{A}^T(k) + \bar{B}_v(k) \cdot R_v \cdot \bar{B}_v^T(k). \end{aligned} \quad (5.12)$$

The reconstructed input vector is thus:

$$\hat{u}(k) = \begin{bmatrix} 1 \\ 1 \\ \delta u_{SOI}(k) \\ \delta \tau_{inj}(k) \\ \delta \hat{p}_{rail}(k) \\ \delta \hat{m}_{air}(k) \\ \delta \hat{p}_{boost}(k) \\ \delta T_e(k) \end{bmatrix}. \qquad (5.13)$$

The injected fuel quantity is not estimated directly by means of the EKF since it is not an input for the control-oriented models. Instead, the input "rail pressure" is chosen as a representative value for the injected fuel quantity and thus it has to be estimated. The corresponding value for the quantity of fuel injected depends only on the inputs "injection duration" and "rail pressure," and is computed on the basis of the following linear model:

$$\delta \hat{V}_{inj}(k) = \theta_{1I}(k) \cdot \delta \tau_{inj}(k) + \theta_{2I}(k) \cdot \delta \hat{p}_{rail}(k). \qquad (5.14)$$

This injection model is structured like the λ and NO_x models presented above in Eqs. (4.3) and (4.4). The parameters θ_{1I} and θ_{2I}, denoting the sensitivities of the injected fuel quantity to injection duration and rail pressure, are derived from the combustion model according to Eqs. (4.17) and (4.18), and are thus defined as:

$$\theta_{1I}(k) = -\theta_{1\lambda}(k) \qquad (5.15)$$
$$\theta_{2I}(k) = -\theta_{2\lambda}(k). \qquad (5.16)$$

The three reconstructed engine inputs "injected fuel quantity," "air mass flow," and "boost pressure" are shown exemplarily in Fig. 5.4 for a fault of the air mass flow.

Figure 5.4 also shows a potential drawback of this FDI strategy: As mentioned above, in the ideal case the EKF estimate of the two non-faulty engine inputs should be zero, whereas the one of the faulty one

5.3. FDI Strategy A

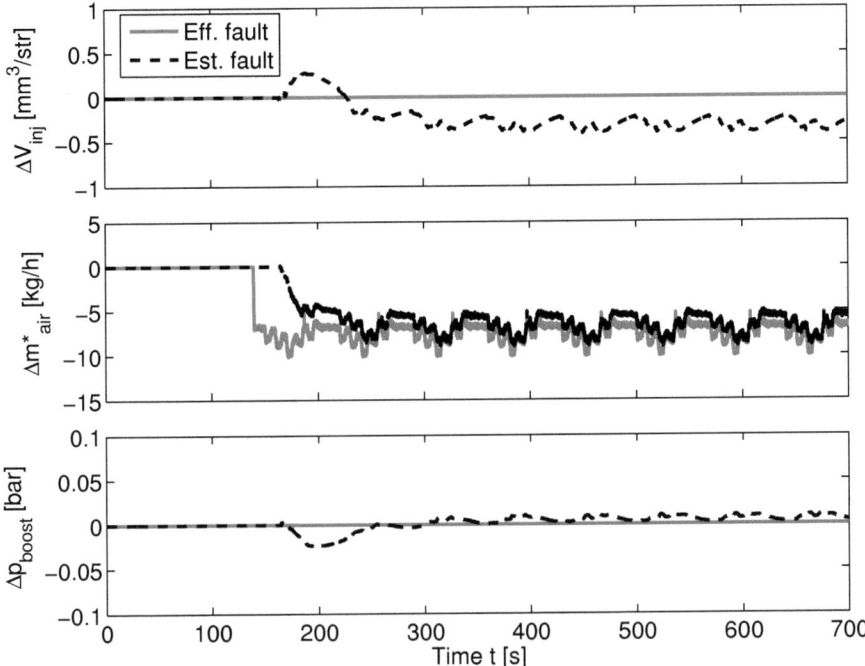

Figure 5.4: FDI strategy A, fault of the air mass flow of -10% introduced at 140 s: Reconstructed injected fuel quantity (top), air mass flow (middle), and boost pressure (bottom).

should be clearly unequal to zero. However, in reality, due to modeling uncertainties, the estimate of the two non-faulty engine inputs are very small but non-zero. Hence the estimation accuracy of the effective faulty input is reduced. Further experimental results to clarify this point are reported in Chapter 6.

5.4 FDI Strategy B: Three Independent EKFs

To improve the fault estimation accuracy the "FDI strategy B" is proposed. In this strategy the λ and NO_x measurements are used to reconstruct the engine inputs suspected to be faulty by inverting the control-oriented models. This is done by means of a bank of EKFs, one for each fault considered. A similar approach was also used in [105] and [111]. Unlike the FDI strategy A, the fault estimation and classification processes are not executed simultaneously, but rather sequentially.

5.4.1 Fault Estimation

A bank of EKFs is developed for the independent estimation of the three engine inputs "injected fuel quantity," "air mass flow," and "boost pressure." Each EKF provides its own quantitative interpretation about the fault status of the engine, as shown in Fig. 5.5, and is specified according to the algorithm of Eqs. (5.2)-(5.12), which has been presented for the FDI strategy A already. Only the faulty input vector $\bar{u}$, the extended state vector $\bar{x}$, and the two system matrices $\bar{A}$ and $\bar{C}$ are different, and thus have to be determined for each of the three EKFs individually. This is done in the following.

5.4. FDI Strategy B

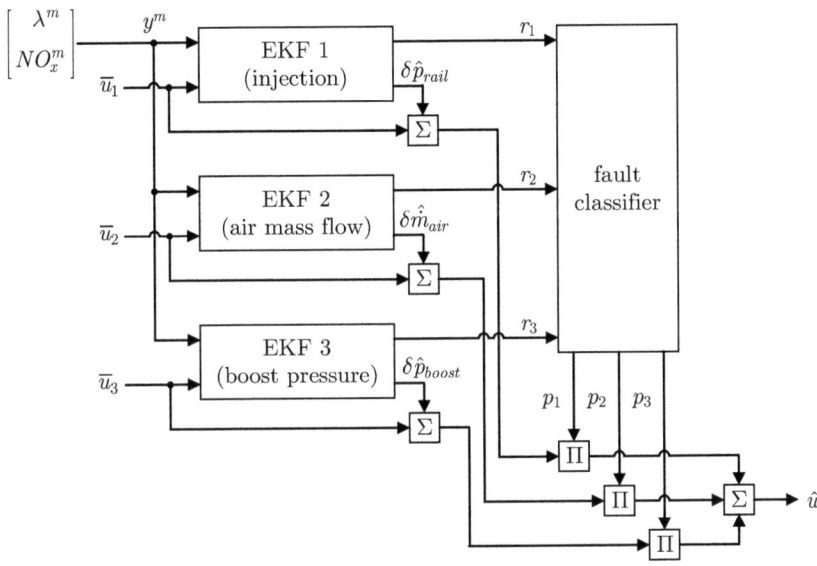

Figure 5.5: The proposed FDI strategy B, based on a bank of three EKFs for the independent estimation of the engine inputs "injected fuel quantity," "air mass flow," and "boost pressure", and a fault classifier.

Injection Quantity

Faulty input and extended state vectors:

$$\bar{u}_1(k) = \begin{bmatrix} 1 \\ 1 \\ \delta u_{SOI}(k) \\ \delta \tau_{inj}(k) \\ 0 \\ \delta \dot{m}_{air}(k) \\ \delta p_{boost}(k) \\ \delta T_e(k) \end{bmatrix}, \quad \bar{x}_1(k) = \begin{bmatrix} \lambda(k) \\ NO_x(k) \\ \delta p_{rail}(k) \end{bmatrix}. \tag{5.17}$$

Extended system matrices:

$$A_{u,1}(k) = \begin{bmatrix} \frac{\partial f_\lambda}{\partial p_{rail}} \\ \frac{\partial f_N}{\partial p_{rail}} \end{bmatrix} = \tilde{B}^*(k) \cdot \begin{bmatrix} \theta_{2\lambda}(k) \\ \theta_{3N}(k) \end{bmatrix}, \qquad (5.18)$$

$$C_{u,1} = \begin{bmatrix} \frac{\partial g_\lambda}{\partial p_{rail}} \\ \frac{\partial g_N}{\partial p_{rail}} \end{bmatrix} = \begin{bmatrix} 0 \\ 0 \end{bmatrix}. \qquad (5.19)$$

Air Mass Flow

Faulty input and extended state vectors:

$$\bar{u}_2(k) = \begin{bmatrix} 1 \\ 1 \\ \delta u_{SOI}(k) \\ \delta \tau_{inj}(k) \\ \delta p_{rail}(k) \\ 0 \\ \delta p_{boost}(k) \\ \delta T_e(k) \end{bmatrix}, \quad \bar{x}_2(k) = \begin{bmatrix} \lambda(k) \\ NO_x(k) \\ \delta \dot{m}_{air}(k) \end{bmatrix}. \qquad (5.20)$$

Extended system matrices:

$$A_{u,2}(k) = \begin{bmatrix} \frac{\partial f_\lambda}{\partial \dot{m}_{air}} \\ \frac{\partial f_N}{\partial \dot{m}_{air}} \end{bmatrix} = \tilde{B}^*(k) \cdot \begin{bmatrix} \theta_{3\lambda}(k) \\ \theta_{4N}(k) \end{bmatrix}, \qquad (5.21)$$

$$C_{u,2} = \begin{bmatrix} \frac{\partial g_\lambda}{\partial \dot{m}_{air}} \\ \frac{\partial g_N}{\partial \dot{m}_{air}} \end{bmatrix} = \begin{bmatrix} 0 \\ 0 \end{bmatrix}. \qquad (5.22)$$

5.4. FDI Strategy B

Boost Pressure

Faulty input and extended state vectors:

$$\bar{u}_3(k) = \begin{bmatrix} 1 \\ 1 \\ \delta u_{SOI}(k) \\ \delta \tau_{inj}(k) \\ \delta p_{rail}(k) \\ \delta \dot{m}_{air}(k) \\ 0 \\ \delta T_e(k) \end{bmatrix}, \quad \bar{x}_3(k) = \begin{bmatrix} \lambda(k) \\ NO_x(k) \\ \delta p_{boost}(k) \end{bmatrix}. \tag{5.23}$$

Extended system matrices:

$$A_{u,3}(k) = \begin{bmatrix} \frac{\partial f_\lambda}{\partial p_{boost}} \\ \frac{\partial f_N}{\partial p_{boost}} \end{bmatrix} = \tilde{B}^*(k) \cdot \begin{bmatrix} 0 \\ \theta_{5N}(k) \end{bmatrix}, \tag{5.24}$$

$$C_{u,3} = \begin{bmatrix} \frac{\partial g_\lambda}{\partial p_{boost}} \\ \frac{\partial g_N}{\partial p_{boost}} \end{bmatrix} = \begin{bmatrix} 0 \\ 0 \end{bmatrix}. \tag{5.25}$$

5.4.2 Fault Classification

Successively, the estimation residuals of each EKF are evaluated to classify the fault. Each EKF $i = 1, 2, 3$ provides also a residual vector r_i for λ and NO_x that could be interpreted as a performance index for the estimation, as shown in Fig. 5.5:

$$r_i(k) = \begin{bmatrix} r_{i,\lambda}(k) \\ r_{i,N}(k) \end{bmatrix} = \begin{bmatrix} \lambda_i^m(k) - \hat{\lambda}_i(k) \\ NO_{x,i}^m(k) - \widehat{NO}_{x,i}(k) \end{bmatrix}. \tag{5.26}$$

The EKF providing the smallest residuals is also the EKF that can best reproduce the fault status of the engine. Following this logic, the classification algorithm is implemented. First, the λ and NO_x residuals of each EKF $i = 1, 2, 3$ are normalized with the corresponding measurements, see Fig. 5.6, and the resulting absolute values are added together, building the combined residuals e_i:

$$e_i(k) = \left|\frac{\lambda_i^m(k) - \hat{\lambda}_i(k)}{\lambda_i^m(k)}\right| + \left|\frac{NO_{x,i}^m(k) - \widehat{NO}_{x,i}(k)}{NO_{x,i}^m(k)}\right|. \quad (5.27)$$

The normalization is necessary to combine the λ and NO_x residuals, since the λ value and the NO_x concentration have different physical dimensions. Successively, the combined residual e_i are cumulated over time, starting from the fault detection time k_{FD}, yielding the cumulated combined residuals ϵ_i, see Fig. 5.7:

$$\epsilon_i = \sum_{k_{FD}}^{k} e_i(k), \quad i = 1, 2, 3. \quad (5.28)$$

At each time step, the smallest cumulated combined residual ϵ_{min} is found, thus indicating the EKF that best matches the actual fault status of the engine:

$$\epsilon_{min}(k) = \min\{\epsilon_1(k), \epsilon_2(k), \epsilon_3(k)\}. \quad (5.29)$$

The three binary factors p_i are computed as follows:

$$[p_1(k), p_2(k), p_3(k)] = \begin{cases} [1, 0, 0] & \text{if } \epsilon_{min}(k) = \epsilon_1(k) \\ [0, 1, 0] & \text{if } \epsilon_{min}(k) = \epsilon_2(k) \\ [0, 0, 1] & \text{if } \epsilon_{min}(k) = \epsilon_3(k) \end{cases}, \quad (5.30)$$

5.4. FDI Strategy B

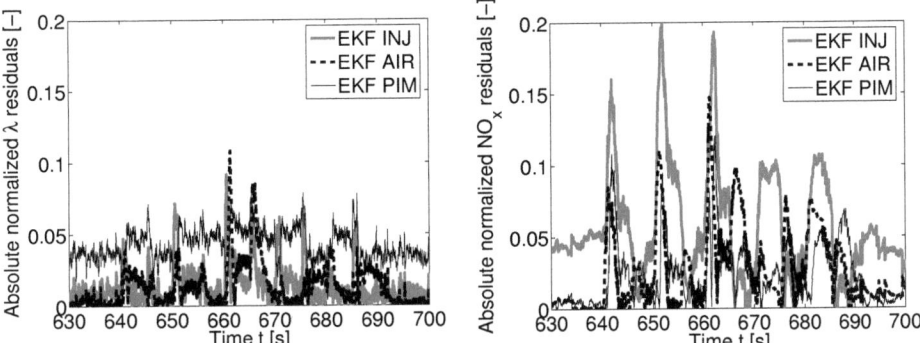

Figure 5.6: Absolute normalized residuals of λ (left) and of NO_x (right). These signals are added together to build the combined residuals of each of the three EKFs.

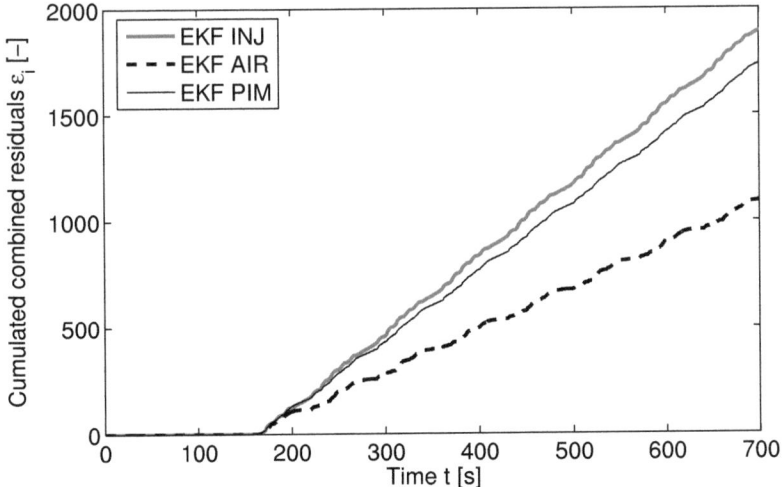

Figure 5.7: Cumulated combined residuals of the three EKFs. The curve with the smallest value indicates the actual fault mode present.

and thus the reconstructed input vector is:

$$\hat{u}(k) = \begin{bmatrix} 1 \\ 1 \\ \delta u_{SOI}(k) \\ \delta \tau_{inj}(k) \\ \delta \hat{p}_{rail}(k) \\ \delta \dot{m}_{air}(k) \\ \delta p_{boost}(k) \\ \delta T_e(k) \end{bmatrix} \cdot p_1(k) + \begin{bmatrix} 1 \\ 1 \\ \delta u_{SOI}(k) \\ \delta \tau_{inj}(k) \\ \delta p_{rail}(k) \\ \delta \hat{\dot{m}}_{air}(k) \\ \delta p_{boost}(k) \\ \delta T_e(k) \end{bmatrix} \cdot p_2(k) + \begin{bmatrix} 1 \\ 1 \\ \delta u_{SOI}(k) \\ \delta \tau_{inj}(k) \\ \delta p_{rail}(k) \\ \delta \dot{m}_{air}(k) \\ \delta \hat{p}_{boost}(k) \\ \delta T_e(k) \end{bmatrix} \cdot p_3(k).$$

(5.31)

Again, the estimated injected fuel quantity is computed according to Eq. (5.14).

The three reconstructed engine inputs "injected fuel quantity," "air mass flow," and "boost pressure" are shown exemplarily in Fig. 5.8 for a fault of the air mass flow.

As confirmed by the comparison between Figs. 5.4 and 5.8, the estimation accuracy is improved. However, more calculation steps are involved in the FDI strategy B. Again, further experimental results are reported in Chapter 6.

5.4. FDI Strategy B

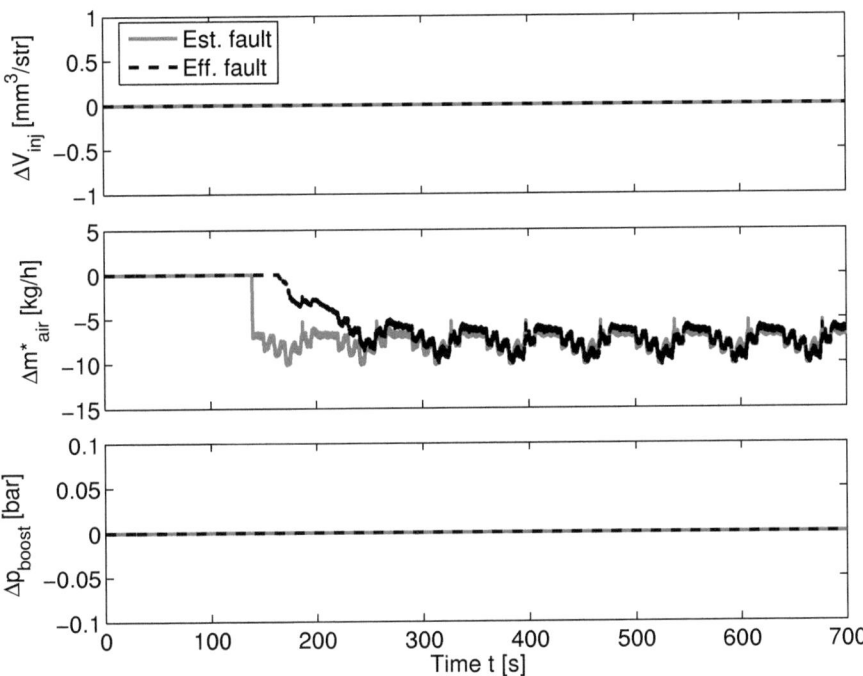

Figure 5.8: FDI strategy B, fault of the air mass flow of -10% introduced at 140 s: Reconstructed injected fuel quantity (top), air mass flow (middle), and boost pressure (bottom).

Chapter 6
Test on the NEDC

The proposed control-oriented emission models and the algorithms for the detection and isolation of faults are tested by means of measurement data of the New European Driving Cycle recorded at the test-bench. The aim is to demonstrate that the control-oriented models are sufficiently accurate to be used for diagnosis purposes, and that the FDI systems can detect, estimate, and classify the various faults considered. The results obtained are evaluated and discussed.

6.1 Measurements

This section explains the procedure used to collect the records of the NEDC at the test-bench.

6.1.1 The Driving Cycle

The control-oriented emission models and the FDI system are verified using measurement data recorded during the NEDC. Prior to performing all measurements, the dynamic brake of the test-bench is programmed to follow the engine speed and torque profiles of the NEDC, using the results of a quasi-static simulation obtained with a dedicated software (QSS-ToolBox, [5]). This procedure is schematically shown in Fig. 6.1. The inputs for the QSS-ToolBox are vehicle speed profile and vehicle data such as weight, aerodynamic coefficients, and gear ratios. The outputs are engine speed profile, torque profile, and fuel consumption. The data of the test vehicle are summarized in Table 6.1.

Table 6.1: Technical data of the vehicle used for the quasi-static simulation.

empty weight		1690 kg
resistance coefficients	aerodynamics	Cx = 0.31
	roll	Cr = 0.011
gear ratios	1st gear	5.30
	2nd gear	3.00
	3rd gear	1.30
	4th gear	1.05
	5ht gear	0.87
	shaft	2.65

6.1.2 Inverse λ and PM Emissions

According to the results shown in Fig. 2.4 of Section 2.3, the cycle-cumulated PM emissions are proportional to the time-averaged value of the inverse λ, but only if $\lambda > 1.2$. Thus the inverse λ can be considered as a "rough indicator" of the PM emissions. For this reason all results concerning the λ model shown in this chapter are based on its inverse value.

6.1.3 Fault Emulation

All faults are introduced synthetically with ASCET, a dedicated software to bypass the ECU of the engine, see Fig. 6.1. The faults

6.1. Measurements

of the injected fuel quantity are introduced by varying the injection duration of -25 μs and -50 μs, causing a variation of the injected fuel volume of about -0.75 mm^3/str and -1.5 mm^3/str, respectively. The faults of the air mass flow are introduced by varying the sensor value by -10%, -5%, $+5\%$ and $+10\%$. The faults of the boost pressure are introduced by varying the set point value by $+5\%$ and $+10\%$.

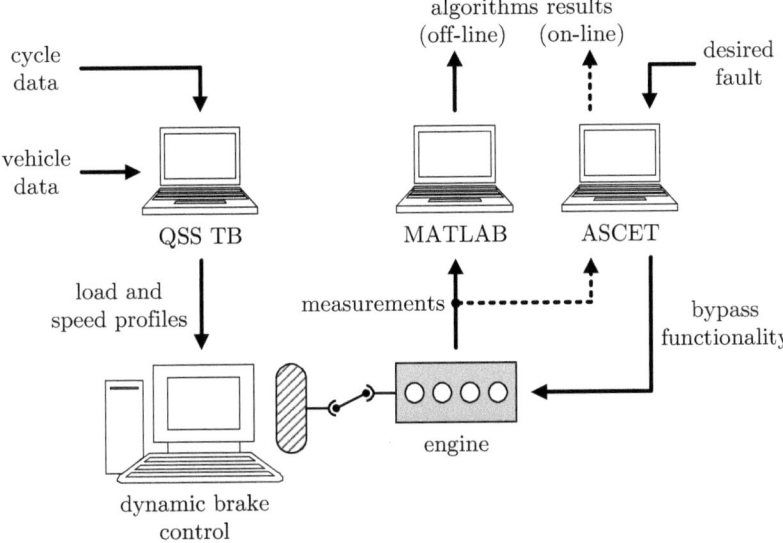

Figure 6.1: The various components of the test-bench and their interactions.

To ensure the same conditions for all measurements, the torque and the engine speed profiles of the dynamic brake are controlled in order to always match those of the NEDC. For example, decreasing the injected fuel quantity causes the engine to decrease the torque produced. At the same time, the test-bench controller increases the gas pedal angle and thus the desired injected fuel volume until the torque prescribed by the NEDC is recovered.

Results obtained from NEDC measurements with all the fault modes described above are shown in Fig. 6.2. The trade-off between NO_x

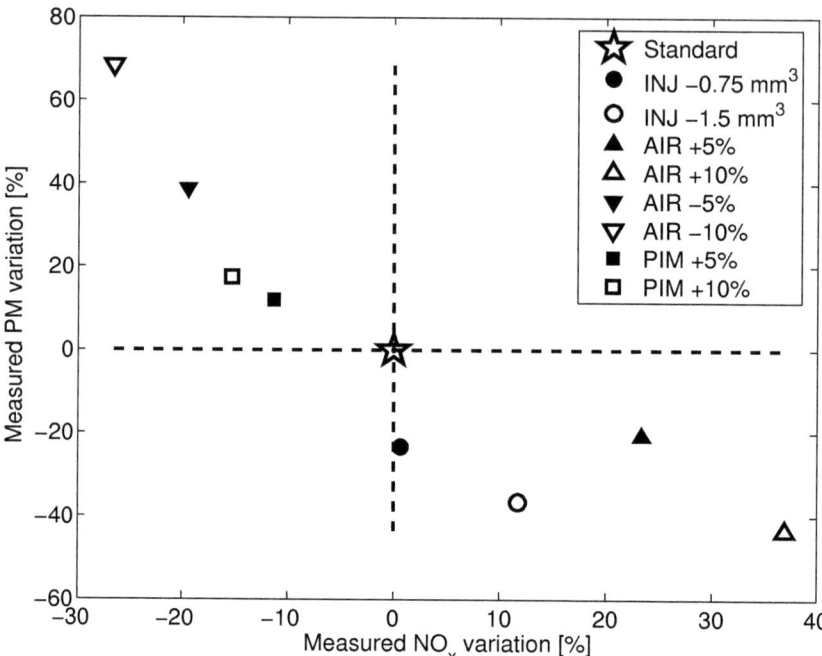

Figure 6.2: Measured cumulated emissions of the fault modes considered. The NO_x are measured with the solid-state sensor and the PM with the PASS.

and PM emissions is evident.

Obviously, the measurement signals of the engine inputs for the fault detection and isolation algorithm should not contain any information about the actual fault introduced. Rather, this information is carried only by the λ and the NO_x sensors. For example, when introducing a fault of the air mass flow sensor of $+5\%$ the corresponding measurement signal used by the fault detection and isolation algorithm should be corrected by -5% in order to keep it as it would be in the fault-free case.

The following experiment is performed for each fault case considered to test the FDI systems. The NEDC is measured once for the

6.1. Measurements

standard fault-free engine, subsequently the fault is introduced and the NEDC is measured for another two consecutive times (Fig. 6.3). The fault introduction is thus equal to 1180 s, which corresponds to the duration of one single measured NEDC.

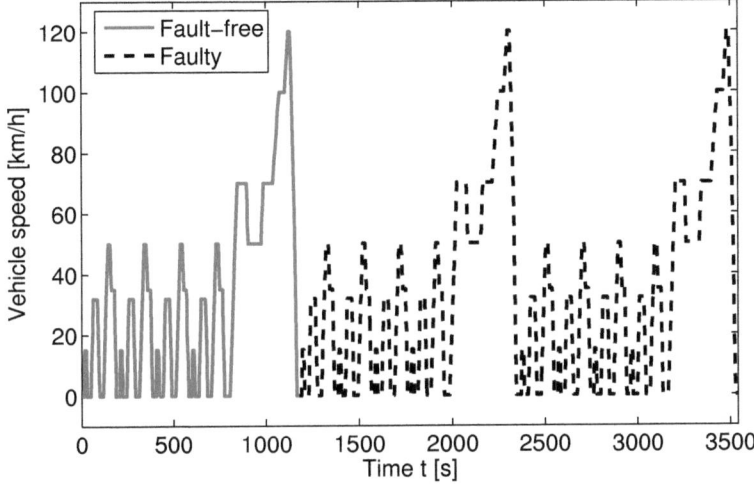

Figure 6.3: The experiment used to test the FDI systems.

A possible application of this work is the detection and isolation of faults due to the ageing of engine parts. Such faults tend to occur after many months of engine operation. Since it is impossible to reproduce such conditions at the test-bench, the faults are introduced artificially as step changes. For the FDI systems of this work, a step change would be the same as a gradual drift: As explained above, a fault is detected once the λ and/or the NO_x residuals exceed certain thresholds, while the way these residuals have grown, i.e., "suddenly" in the case of a step change or "gradually" in the case of a drift, is not relevant for the decision about the fault detection nor for the subsequent estimation and classification of the fault.

6.2 Results

This section shows the results obtained by the proposed control-oriented models and the two FDI strategies on the basis of measurement data of the NEDC recorded at the test-bench.

6.2.1 Control-Oriented Emission Models

Figure 6.4 shows the comparison between instantaneous measured and simulated NO_x emission for the standard fault-free engine during the last half of the NEDC, including one urban and the highway cycle. Analogously, Fig. 6.5 shows the comparison between instantaneous measured and simulated value of inverse λ. Figure 6.6 shows the comparison between measured and simulated cumulated NO_x emission for all performed measurements of the NEDC, i.e. standard fault-free and faulty. Analogously, Fig. 6.7 shows the comparison between averaged inverse measured and simulated λ. Finally, Fig. 6.8 shows the comparison between measured and simulated cumulated NO_x emission during the NEDC obtained from data of vehicles of different weights. Here, results from a simple reference-based NO_x model, i.e, a static map of the NO_x concentration, are also shown[1].

These results clearly show that both control-oriented models can well reproduce the characteristics of the standard fault-free engine, as well as those of the faulty engine. This fact is a fundamental prerequisite for the development of diagnostic algorithms.

According to the idea illustrated by Fig. 4.9, prior to be used for diagnostics purposes, the NO_x model is adapted on the basis of real NEDC measurements. The adaptation algorithm is activated, and the NEDC is repeated until the accuracy of the NO_x prediction is satisfactory. The results of this procedure are shown in Fig. 6.9. Even when non-adapted parameters are used, the accuracy of the model is clearly higher than those obtained with the simple reference-based

---

[1]The engine speed and torque profiles used here have not been derived as described in Section 6.1.1, but are available from other measurement data of vehicles with the same engine as the one used for this work.

6.2. Results

NO_x model, i.e, a static map of the NO_x concentration. By repeating the NEDC during the adaptation, the simulated NO_x emission converges to the measured value. However, after eight repetitions, the accuracy no longer improves significantly, and the adaptation can be deactivated.

6.2.2 FDI Strategy A

The results from the FDI strategy A are shown. Figures 6.10 and 6.11 show results from faults of the injected fuel quantity, Figs. 6.12-6.15 from faults of the air mass flow sensor, and Figs. 6.16 and 6.17 from faults of the boost pressure sensor. To clarify the effect on the FDI system of the adaptation procedure described in Section 4.3, results obtained with adapted parameters (AP) and with non-adapted parameters (nAP) of the NO_x model are shown.

6.2.3 FDI Strategy B

The results from the FDI strategy B are shown. Figures 6.18 and 6.19 show results from faults of the injected fuel quantity, Figs. 6.20-6.23 from faults of the air mass flow sensor, and Figs. 6.24 and 6.25 from faults of the boost pressure sensor. Also in this case, but only where evident, results obtained with adapted parameters (AP) and with non-adapted parameters (nAP) of the NO_x model are shown in order to clarify the effect on the FDI system of the adaptation procedure described in Section 4.3.

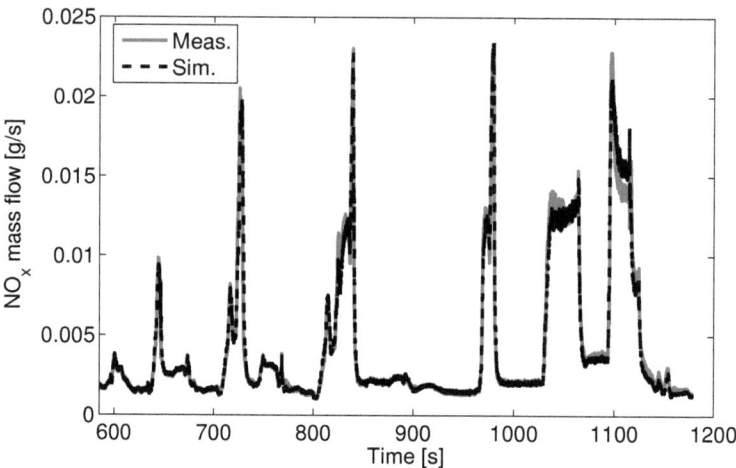

Figure 6.4: Measured and simulated NO_x emission for the standard fault-free engine during the last half of the NEDC.

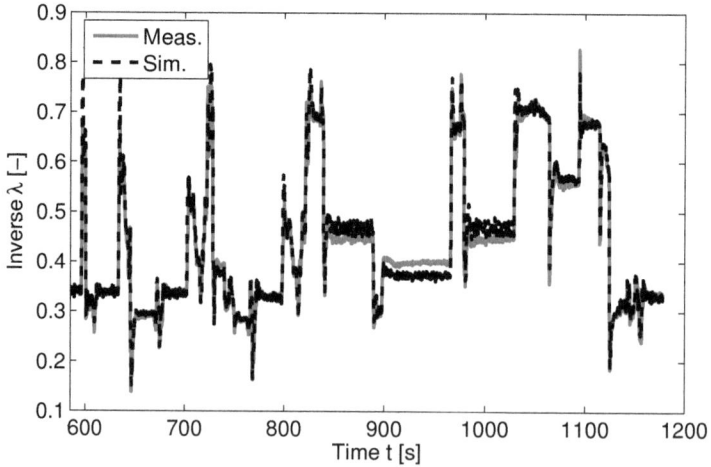

Figure 6.5: Measured and simulated inverse λ for the standard fault-free engine during the last half of the NEDC.

6.2. Results

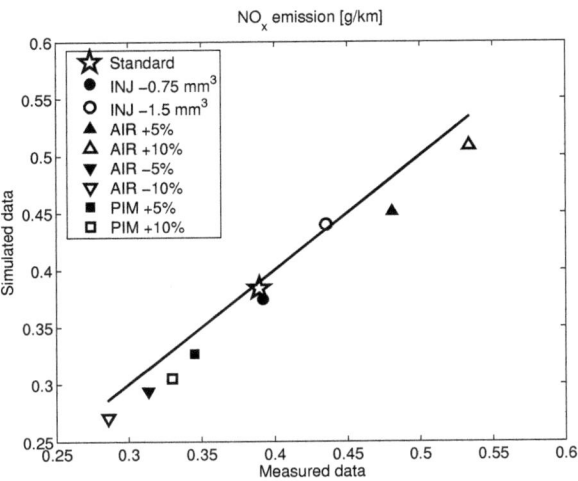

Figure 6.6: Measured vs. simulated cumulated NO_x emission for all NEDC measurements, i.e., standard fault-free and faulty.

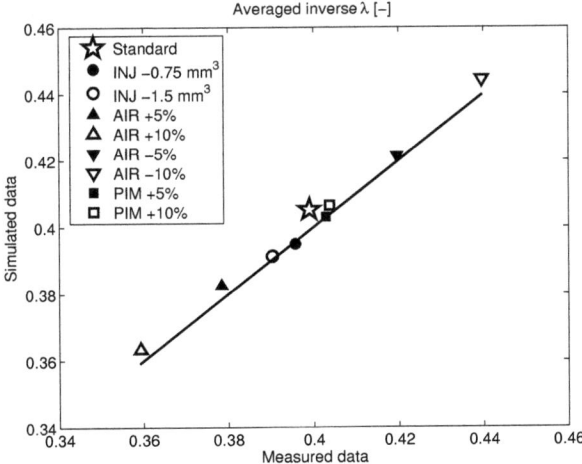

Figure 6.7: Measured vs. simulated averaged inverse λ for all NEDC measurements, i.e., standard fault-free and faulty.

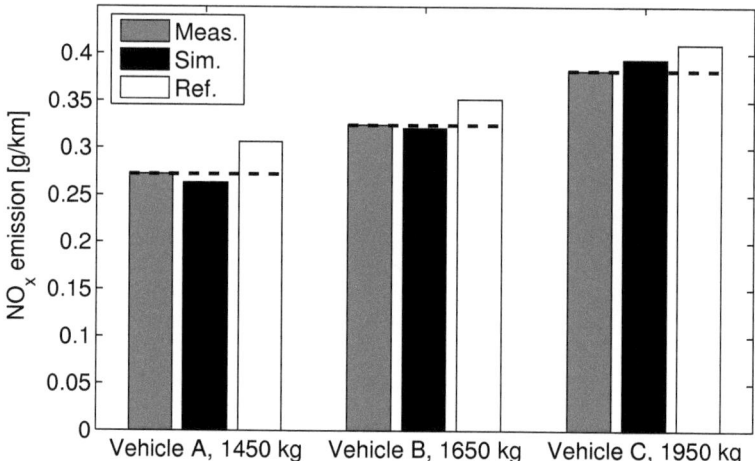

Figure 6.8: Comparison between measured, reference-based, and simulated cumulated NO_x emission during the NEDC for vehicles with different weights.

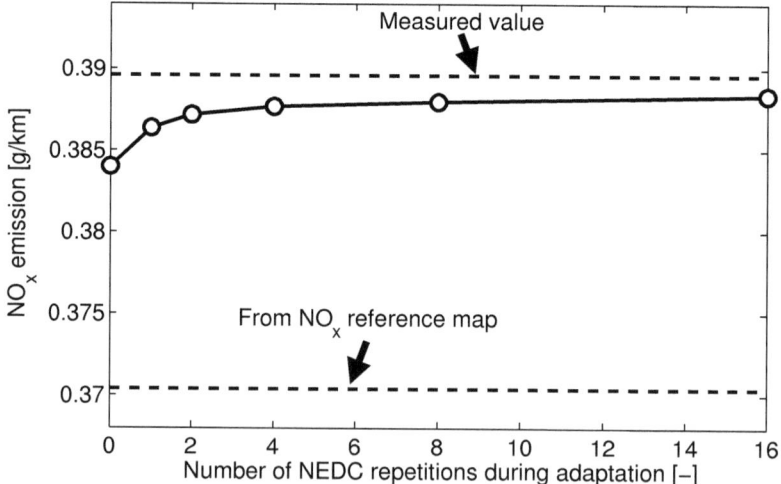

Figure 6.9: Effect of the number of NEDC repetitions during the adaptation on the prediction quality of the simulated NO_x emission.

6.2. Results

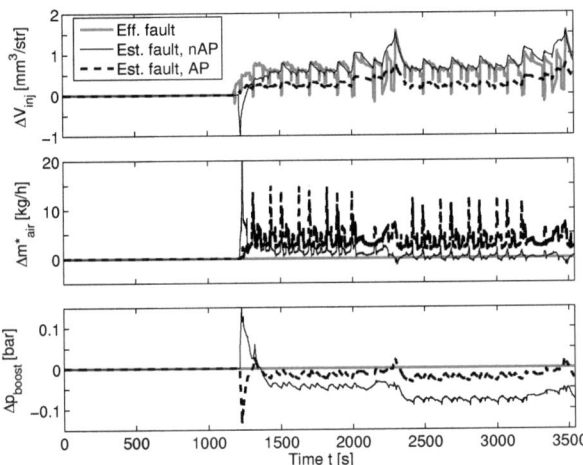

Figure 6.10: Results of the FDI system, strategy A, for a fault of the injected fuel quantity of $-0.75\ mm^3/str$ introduced at 1180 s.

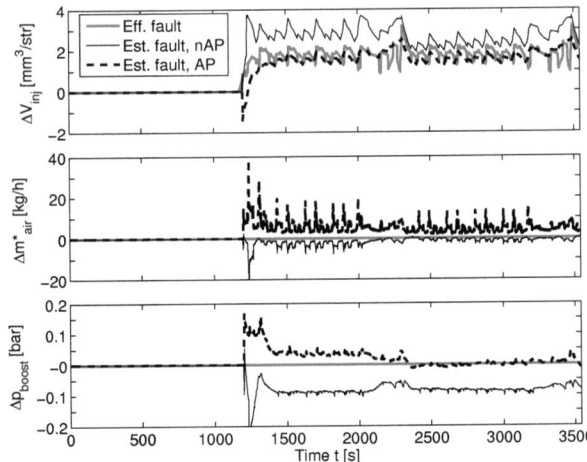

Figure 6.11: Results of the FDI system, strategy A, for a fault of the injected fuel quantity of $-1.5\ mm^3/str$ introduced at 1180 s.

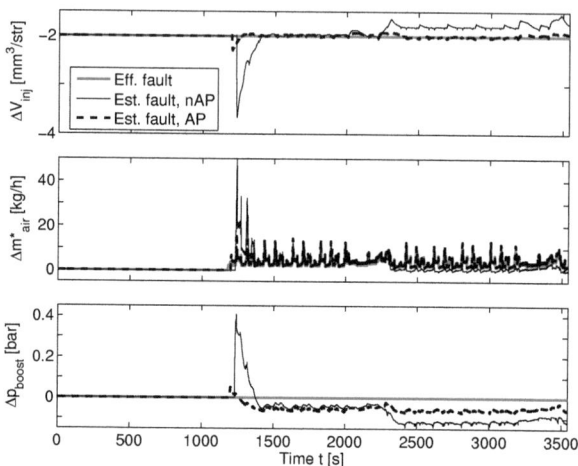

Figure 6.12: Results of the FDI system, strategy A, for a fault of the air mass flow of $+5\%$ introduced at $1180\ s$.

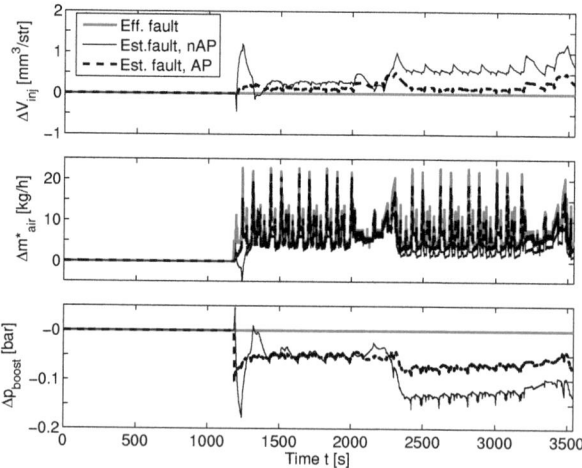

Figure 6.13: Results of the FDI system, strategy A, for a fault of the air mass flow of $+10\%$ introduced at $1180\ s$.

6.2. Results

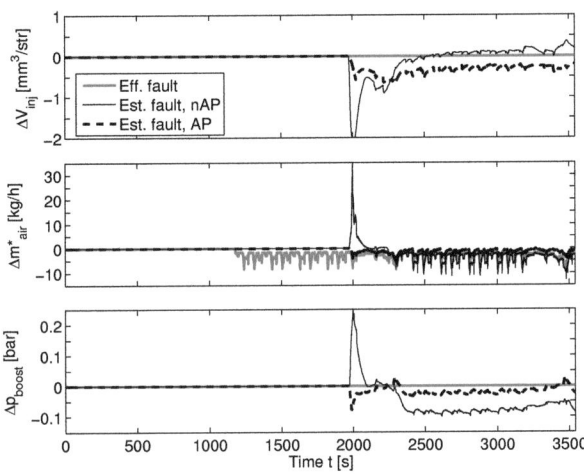

Figure 6.14: Results of the FDI system, strategy A, for a fault of the air mass flow of -5% introduced at $1180\ s$.

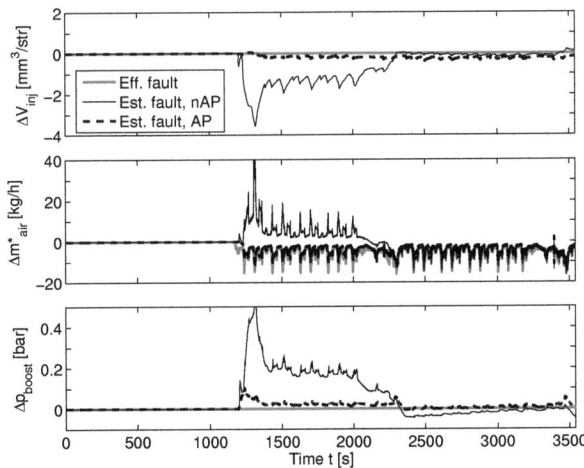

Figure 6.15: Results of the FDI system, strategy A, for a fault of the air mass flow of -10% introduced at $1180\ s$.

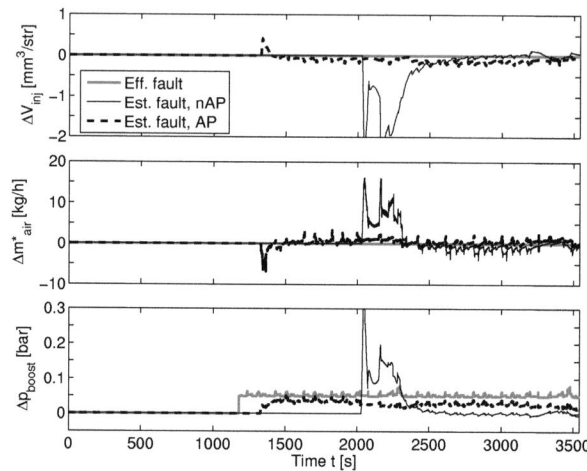

Figure 6.16: Results of the FDI system, strategy A, for a fault of the boost pressure of +5% introduced at 1180 s.

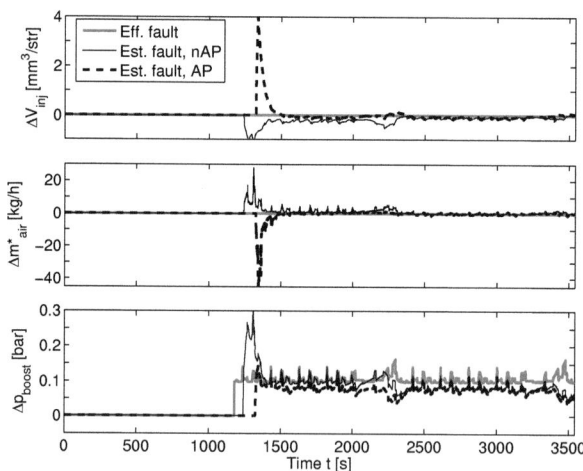

Figure 6.17: Results of the FDI system, strategy A, for a fault of the boost pressure of +10% introduced at 1180 s.

6.2. Results

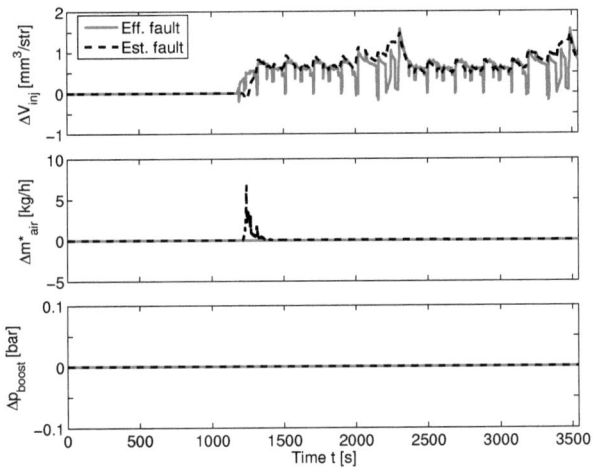

Figure 6.18: Results of the FDI system, strategy B, for a fault of the injected fuel quantity of $-0.75\ mm^3/str$ introduced at 1180 s.

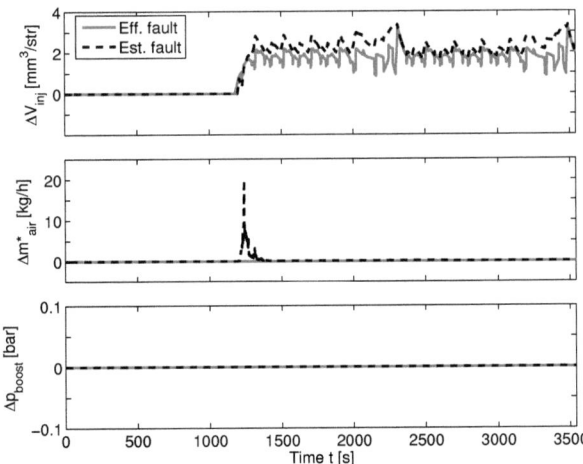

Figure 6.19: Results of the FDI system, strategy B, for a fault of the injected fuel quantity of $-1.5\ mm^3/str$ introduced at 1180 s.

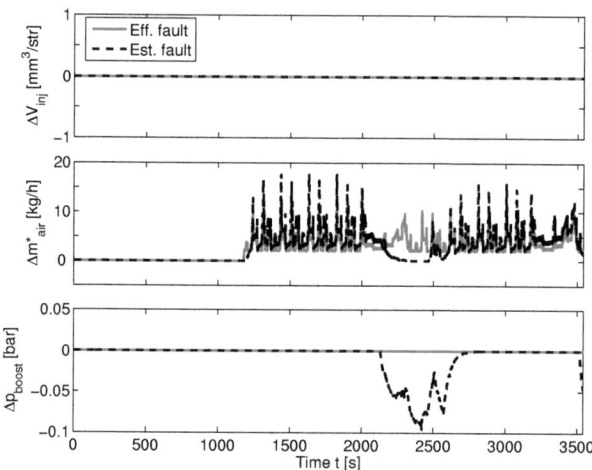

Figure 6.20: Results of the FDI system, strategy B, for a fault of the air mass flow of +5% introduced at 1180 s.

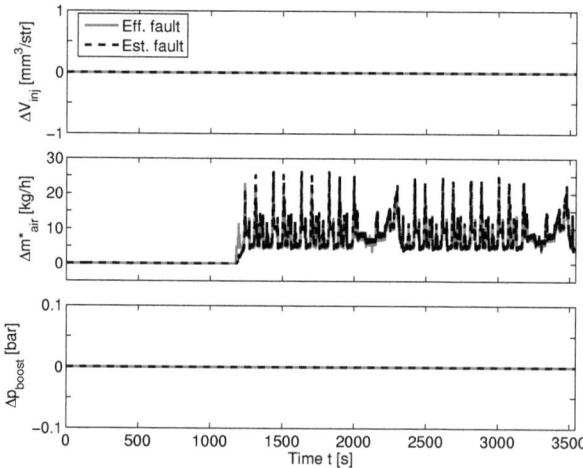

Figure 6.21: Results of the FDI system, strategy B, for a fault of the air mass flow of +10% introduced at 1180 s.

6.2. Results

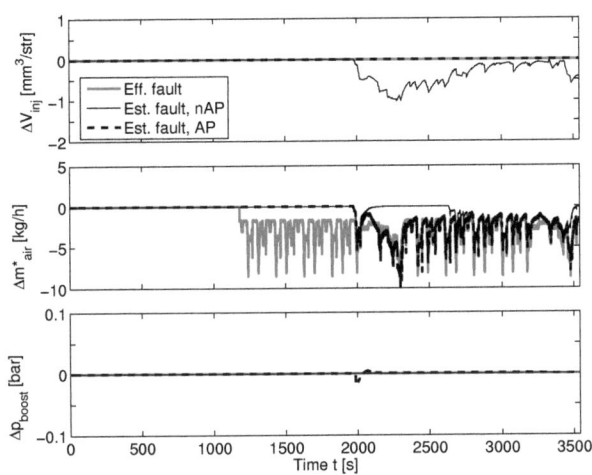

Figure 6.22: Results of the FDI system, strategy B, for a fault of the air mass flow of -5% introduced at $1180\ s$.

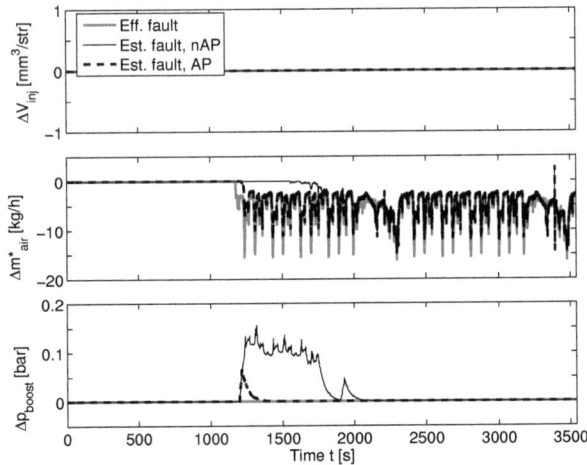

Figure 6.23: Results of the FDI system, strategy B, for a fault of the air mass flow of -10% introduced at $1180\ s$.

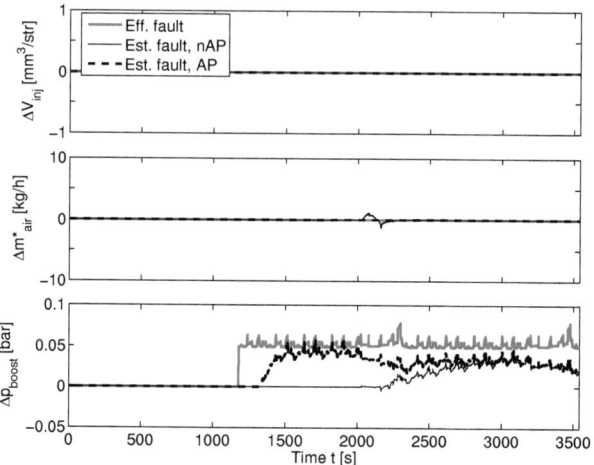

Figure 6.24: Results of the FDI system, strategy B, for a fault of the boost pressure of +5% introduced at 1180 s.

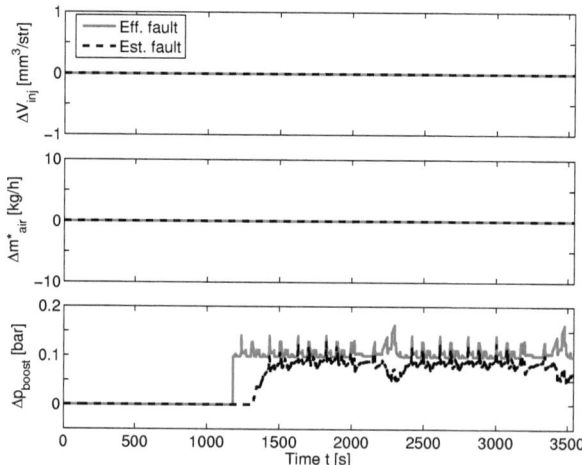

Figure 6.25: Results of the FDI system, strategy B, for a fault of the boost pressure of +10% introduced at 1180 s.

6.3 Discussion

Fault Detection

The first point to be analyzed is the fault detection. The performance index to evaluate the quality of the fault detection is represented by the time needed to detect the faults, denoted as "fault detection time" (FDT). Ideally, the FDT should be zero, indicating an instantaneous detection of the fault. Table 6.2 summarizes the FDT for all fault modes considered as well as the effect of the adaptation procedure.

Table 6.2: Time needed to detect the fault (FDT = fault detection time) for both strategies A and B.

fault introduced	FDT [s]	effect of adaptation
INJ $-0.75\ mm^3/str$	37	none
INJ $-1.5\ mm^3/str$	10	none
AIR $+5\%$	28	none
AIR $+10\%$	5	none
AIR -5%	793	none
AIR -10%	20	none
PIM $+5\%$	152	FDT $\downarrow$
PIM $+10\%$	144	none

As shown, all faults introduced are detected quite quickly. Only in one case, i.e., "AIR -5%," a considerable FDT is observed. The adaptation procedure plays a marginal role for the quality of the fault detection, and only in one case the quality of the fault detection is improved considerably, i.e., for "PIM $+5\%$" (Figs. 6.16 and 6.24). By using adapted parameters, the FDT is clearly reduced by more than $500\ s$.

Successively, fault estimation and classification are analyzed for both proposed FDI strategies A and B.

FDI Strategy A

For the FDI strategy A, fault estimation and classification are coupled. As mentioned in Section 5.3, in the ideal case the EKF estimate of the two non-faulty engine inputs should be zero, whereas those of the faulty one should clearly not be equal to zero. Unlike for the fault detection, it is not possible to establish a quantitative performance index to evaluate the estimation and classification capabilities of this FDI strategy. Rather, these capabilities are determined directly by observing the results shown in Figs. 6.10-6.17.

The results obtained by using non-adapted parameters of the NO_x model are inadequate, whereas the tendency of the EKF estimate to reach the ideal case described above by using adapted parameters of the NO_x model is clear. Consider for instance the fault mode "AIR +10%" of Fig. 6.13: By using non-adapted parameters, simultaneous faults of the injected fuel quantity, of the air mass flow, and of the boost pressure are erroneously estimated. By using adapted parameters, the estimated fault of the injected fuel quantity is close to zero, the estimated fault of the air mass flow is more consistent with the fault effectively introduced, and the estimated fault of the boost pressure, although it is still non-zero, is clearly smaller. Another typical example is the fault mode "AIR -10%" of Fig. 6.15: Although using both non-adapted and adapted parameters leads to the same correct final result, in the first case the estimation is completely erroneous during about 1000 s after the fault detection, whereas in the second case the estimation is immediately correct.

The benefits of using adapted parameters of the NO_x model are evident in most of the fault modes considered. Thus the adaptation process is a fundamental prerequisite for the proper working of the FDI strategy A.

FDI Strategy B

For the FDI strategy B, fault estimation and classification are two distinct processes, which have to be evaluated independently.

6.3. Discussion

Also, in this case it is not possible to establish a quantitative performance index to evaluate the estimation capabilities of this FDI strategy. Rather, these capabilities are determined directly by observing the results shown in Figs. 6.18-6.25. However, a performance index to evaluate the quality of the fault classification can be established. It is represented by the time during which a given fault mode is erroneously classified as another fault mode, denoted as "erroneous classification time" (ECT). Ideally the ECT should be zero, indicating an instantaneous, correct classification of the fault. Table 6.3 summarizes the results of the the ECT for all fault modes considered and the effect of the adaptation procedure.

Table 6.3: FDI strategy B, classification quality (ECT = erroneous classification time).

fault introduced	ECT [s]	effect of adaptation
INJ $-0.75\ mm^3/str$	148	none
INJ $-1.5\ mm^3/str$	150	none
AIR $+5\%$	0	none
AIR $+10\%$	0	none
AIR -5%	0	ECT $\downarrow$
AIR -10%	17	ECT $\downarrow$
PIM $+5\%$	0	none
PIM $+10\%$	0	none

Surprisingly, for this case the adaptation plays a marginal role. The estimation quality is neither improved nor worsened by using adapted parameters, and only in two cases the classification is improved substantially (Figs. 6.22 and 6.23). This facts indicates that the FDI strategy B is more robust against modeling uncertainties than the FDI strategy A.

In several cases, for instance in those shown in Figs. 6.18 and 6.19, an erroneous fault classification immediately after the fault detection is observed. However, in all of these cases the fault is correctly classified after a certain period of time, and the correct decisions remained unaltered until the end of the experiments. Thus the classification algorithm of the FDI seems to converge to an unequivocal decision. A possible cause of the classification errors observed during the first phase of the FDI algorithm is that the three EKFs are not yet in steady-state conditions.

Chapter 7
Conclusions

Results from experiments conducted at the test-bench engine show that the control-oriented emission models developed can reproduce the emission characteristics of the standard fault-free engine as well as those of the faulty engine. Thus they are applicable for diagnostics purposes. Using the adaptation procedure an accurate NO_x model can be successfully obtained, even if the initial parameter set calculated with the combustion model is uncertain.

Experimental results show that the adaptation improves the accuracy of the NO_x model, match the characteristics of each individual engine, and enhance its transferability to other engines.

Further dynamic measurements recorded at the test-bench show that both FDI strategies developed are able to detect, classify, and estimate all faults considered. For a proper functioning, the FDI strategy A needs accurate models, which clearly makes the adaptation procedure a fundamental prerequisite. Instead, in the case of the FDI strategy B only in two experiments the fault classification capabilities are improved because of adapted parameters. Hence the FDI strategy B is more robust against modeling uncertainties than the FDI strategy A. Furthermore, the fault estimation accuracy of FDI strategy B is higher than FDI strategy A. On the other hand the FDI strategy A presents a simpler structure than FDI strategy B: While in the first case a single EKF is implemented, for the latter a bank of three EKFs with a classification algorithm is necessary.

The following four steps summarize the preferred methodology of this work:

- Off-line step 1: Calculation of the parameter maps for the λ and NO_x models using the combustion model developed (or alternatively a similar tool known by the user).

- Off-line step 2: Write calculated maps of parameters in the ECU of the engines.

- On-line step 1: Activation of the NO_x model parameter adaptation during the normal engine operation to improve the accuracy and to match the characteristics of each individual engine.

- On-line step 2: Deactivation of the NO_x model parameter adaptation and activation of FDI strategy B during the normal engine operation in order to detect and isolate faults of the injected fuel quantity, air mass flow, and boost pressure.

   FDI strategy B is preferred because of its higher fault estimation accuracy and of its robustness against modeling uncertainties.

The results of the presented FDI methodology are time and size of the deviation of quantity of injected fuel, air mass flow, and boost pressure from their nominal values.

The FDI methodology does not indicate the source of these deviations. For example, an air mass flow deviation could be caused either by an air mass flow sensor fault or by an EGR valve fault. Also it cannot classify faults of individual cylinders, like single injector faults, misfires, etc.

Appendix A

Complete Equation and Parameter Set of the Combustion Model

A.1 Injection

- Reference maximum injection rate:

$$\frac{dm^0_{inj,max}}{d\varphi} \approx \frac{1}{\omega_e} \cdot c_d \cdot A_{noz} \cdot \sqrt{2 \cdot \rho_{f,inj} \cdot p^0_{rail}} \quad \text{(A.1)}$$

- Maximum injection rate in another operating point:

$$\frac{dm_{inj,max}}{d\varphi} = \frac{dm^0_{inj,max}}{d\varphi} \cdot \left(\frac{p_{rail}}{p^0_{rail}}\right)^{n_1} \quad \text{(A.2)}$$

- Injection rate slope in another operating point:

$$\alpha = \alpha^0 \cdot \left(\frac{p_{rail}}{p^0_{rail}}\right)^{n_2} \quad \text{(A.3)}$$

- Injector closing is faster than injector opening:

$$\alpha_{up} = \alpha \quad \text{(A.4)}$$
$$\alpha_{down} = c \cdot \alpha_{up} \quad \text{(A.5)}$$

132   Appendix A. Combustion Model: Equations and Parameters

param.	physical meaning	value	unit
c_d	discharge coefficient	0.85	–
d_{noz}	nozzle hole diameter	$169 \cdot 10^{-6}$	m
n_1	empirical parameter	0.52	–
n_2	empirical parameter	1.3	–
c	empirical parameter	1	–
α^0_{pilot}	reference slope of pilot injection rate	32.93	kg/s^2
α^0_{main}	reference slope of main injection rate	45.93	kg/s^2
p^0_{rail}	reference rail pressure	$500 \cdot 10^5$	Pa
τ_{open}	needle opening delay	$300 \cdot 10^{-6}$	s
τ_{close}	needle closing delay	$300 \cdot 10^{-6}$	s

A.2 Heat Release

A.2.1 Evaporation of the Injected Fuel

- Actual injected fuel mass:

$$\Delta m_{inj}(\varphi) = \frac{dm_{inj}}{d\varphi} \cdot \Delta \varphi_{sample} \qquad (A.6)$$

- Number of injected fuel droplets:

$$n_{dr}(\varphi) = \frac{\Delta m_{inj}(\varphi)}{m_{dr,liq}(\varphi)} \qquad (A.7)$$

- Characteristic volume and mass of the injected liquid fuel droplets:

$$V_{dr,liq}(\varphi) = d^3_{dr,liq}(\varphi) \cdot \frac{\pi}{6} \qquad (A.8)$$
$$m_{dr,liq}(\varphi) = \rho_f \cdot V_{dr,liq}(\varphi) \qquad (A.9)$$

- Diameter of the injected liquid fuel droplets (Sauter mean diameter):

$$d_{dr,liq} = c_{SMD} \cdot d_{noz,eff} \cdot (Re \cdot We)^{-0.28} \qquad (A.10)$$

A.2. Heat Release

- Reynolds' and Weber's numbers:

$$Re = \frac{u_{dr,liq} \cdot d_{noz,eff}}{\nu_f} \tag{A.11}$$

$$We = \frac{u_{dr,liq}^2 \cdot d_{noz,eff} \cdot \rho_{cyl}}{\sigma_f} \tag{A.12}$$

- Droplet velocity:

$$u_{dr,liq} = \frac{\frac{dm_{inj}}{dt}}{\rho_f \cdot A_{noz,eff}} \tag{A.13}$$

- Actual diameter of the liquid fuel droplets (injected fuel droplet evaporation, d^2 law):

$$d_{dr}^2 = d_{dr,liq}^2 - \beta \cdot t \tag{A.14}$$

- Characteristic volume and mass of the actual liquid fuel droplets:

$$V_{dr}(\varphi) = d_{dr}^3(\varphi) \cdot \frac{\pi}{6} \tag{A.15}$$

$$m_{dr}(\varphi) = \rho_f \cdot V_{dr}(\varphi) \tag{A.16}$$

- Actual liquid fuel mass:

$$\Delta m_{liq}(\varphi) = m_{dr}(\varphi) \cdot n_{dr} \tag{A.17}$$

- Actual evaporated fuel mass:

$$\Delta m_{vap}(\varphi) = \Delta m_{inj}(\varphi) - \Delta m_{liq}(\varphi) \tag{A.18}$$

- Total evaporated fuel mass:

$$m_{vap,tot}(\varphi) = m_{vap,tot}(\varphi - \Delta\varphi_{sample}) + \Delta m_{vap}(\varphi) \tag{A.19}$$

Appendix A. Combustion Model: Equations and Parameters

- Evaporated fuel mass in one single premixed zone:

$$m_{vap,zone}(\varphi) = \frac{m_{vap,tot}(\varphi)}{n_{noz}} \qquad (A.20)$$

$$\frac{dm_{vap,zone}}{d\varphi} = \frac{m_{vap,zone}(\varphi) - m_{vap,zone}(\varphi - \Delta\varphi_{sample})}{\Delta\varphi_{sample}} \qquad (A.21)$$

- Air and exhaust gas mass in the zone:

$$\frac{dm_{aireg,zone}}{d\varphi} = \Lambda_0 \cdot \chi_{st} \cdot \frac{dm_{vap,zone}}{d\varphi} \qquad (A.22)$$

$$m_{air,zone}(\varphi) = \frac{m_{cyl,air}(\varphi)}{m_{cyl}(\varphi)} \cdot m_{aireg,zone}(\varphi) \qquad (A.23)$$

$$m_{eg,zone}(\varphi) = m_{aireg,zone}(\varphi) - m_{air,zone}(\varphi) \qquad (A.24)$$

- Relative A/F ratios:

$$\lambda_{zone}(\varphi) = \frac{m_{air,zone}(\varphi)}{\chi_{st} \cdot m_{vap,zone}(\varphi)} \qquad (A.25)$$

$$\Lambda_{zone}(\varphi) = \frac{m_{aireg,zone}(\varphi)}{\chi_{st} \cdot m_{vap,zone}(\varphi)} \qquad (A.26)$$

param.	physical meaning	value	unit
d_{noz}	nozzle hole diameter	$169 \cdot 10^{-6}$	m
n_{noz}	number of nozzle holes	6	–
c_{SMD}	Sauter mean diameter scaling factor	4	–
β	empirical evaporation rate	$7 \cdot 10^{-6}$	m^2/s
Λ_0	initial rel. air+exhaust gas to fuel ratio	0.9	–
χ_{st}	stoichiometric A/F ratio	14.67	–
ρ_f	fuel density	824.1	kg/m^3
ν_f	fuel kinematic viscosity	$2.496 \cdot 10^{-6}$	m^2/s
σ_f	fuel surface tension	0.03	N/m

A.2. Heat Release

A.2.2 Air/fuel Mixture Preparation

- Premixed zone mass:
$$m_{zone}(\varphi) = m_{vap,zone}(\varphi) + m_{aireg,zone}(\varphi) \tag{A.27}$$

- Premixed zone density:
$$\rho_{zone}(\varphi) = \frac{m_{vap,zone}(\varphi) \cdot \rho_f + m_{air,zone}(\varphi) \cdot \rho_{air}(T_{cyl}) + m_{eg,zone}(\varphi) \cdot \rho_{eg}}{m_{zone}(\varphi)} \tag{A.28}$$

- Volume of the premixed zone (assumption: spherical):
$$V_{zone}(\varphi) = \frac{m_{zone}(\varphi)}{\rho_{zone}(\varphi)} \tag{A.29}$$

- Diameter of the premixed zone:
$$d_{zone}(\varphi) = \left[\frac{6}{\pi} \cdot V_{zone}(\varphi)\right]^{\frac{1}{3}} \tag{A.30}$$

- Surface of the premixed zone (assumption: spherical):
$$A_{zone}(\varphi) = \frac{\pi}{4} \cdot d_{zone}^2(\varphi) \tag{A.31}$$

- Reynolds' number:
$$Re = \frac{c_m \cdot V_{cyl}^{\frac{1}{3}}(\varphi)}{\nu_{air}(T_{cyl})} \tag{A.32}$$

- Mean piston speed:
$$c_m = \frac{V_d \cdot \omega_e}{\pi} \tag{A.33}$$

- Fuel density of the premixed zone:
$$\rho_{f,zone}(\varphi) = \frac{m_{vap,zone}(\varphi)}{V_{zone}} \tag{A.34}$$

Appendix A. Combustion Model: Equations and Parameters

- Diffusion of fuel out of the premixed zone and air entrainment:

$$\frac{dm_{vap,zone}}{d\varphi} = \frac{1}{\omega_e} \cdot c_1 \cdot Re^{c_2}(\varphi) \cdot A_{zone}(\varphi) \cdot \frac{\rho_{f,zone}(\varphi)}{d_{zone}(\varphi)} \quad (A.35)$$

- Actual diffused fuel mass:

$$\Delta m_{vap,zone}(\varphi) = \frac{dm_{vap,zone}}{d\varphi} \cdot \Delta\varphi_{sample} \quad (A.36)$$

- Fuel-to-air + exhaust gas ratio:

$$\xi(\varphi) = \frac{m_{vap,zone}(\varphi) - \Delta m_{vap,zone}(\varphi)}{m_{aireg}(\varphi)} \quad (A.37)$$

- Air and exhaust gas mass in the zone:

$$m_{aireg,zone}(\varphi + \Delta\varphi_{sample}) = \begin{cases} \frac{m_{vap,zone}(\varphi)}{\xi(\varphi)} & \text{if } \xi > 0 \\ m_{aireg,zone}(\varphi) & \text{if } \xi = 0 \end{cases} \quad (A.38)$$

$$m_{air,zone}(\varphi) = \frac{m_{cyl,air}(\varphi)}{m_{cyl}(\varphi)} \cdot m_{aireg,zone}(\varphi) \quad (A.39)$$

$$m_{eg,zone}(\varphi) = m_{aireg,zone}(\varphi) - m_{air,zone}(\varphi) \quad (A.40)$$

- Relative A/F ratios:

$$\lambda_{zone}(\varphi) = \frac{m_{air,zone}(\varphi)}{X_{st} \cdot m_{vap,zone}(\varphi)} \quad (A.41)$$

$$\Lambda_{zone}(\varphi) = \frac{m_{aireg,zone}(\varphi)}{X_{st} \cdot m_{vap,zone}(\varphi)} \quad (A.42)$$

param.	physical meaning	value	unit
V_d	cylinder displacement	$0.538 \cdot 10^{-3}$	m^3
X_{st}	stoichiometric A/F ratio	14.67	—
ρ_f	fuel density	824.1	kg/m^3
c_1	diffusion-caused dilution scaling factor	$2 \cdot 10^{-4}$	m^2/s
c_2	Reynolds' number exponential factor	0.6	—

A.2. Heat Release

A.2.3 Ignition Delay

- Total ignition delay:

$$\tau_{ID}(\varphi) = \tau_{phys} + \tau_{chem}(\varphi) \tag{A.43}$$

- Physical part:

$$\tau_{phys} = c_1 \cdot u_{dr,liq}^{-1.68} \cdot d_{noz,eff}^{0.88} \tag{A.44}$$

- Chemical part:

$$\tau_{chem}(\varphi) = c_2 \cdot \left(\frac{p_{cyl}(\varphi)}{p_{ref}}\right)^{c_3} \cdot \lambda_{zone}^{c_4}(\varphi) \cdot e^{\frac{T_A}{T_{cyl}(\varphi)}} \tag{A.45}$$

- Condition for ignition:

$$\int_{SOI}^{SOC} \frac{1}{\tau_{ID}(\varphi) \cdot \omega_e} d\varphi \geq 1 \tag{A.46}$$

param.	physical meaning	value	unit
d_{noz}	nozzle hole diameter	$169 \cdot 10^{-6}$	m
p_{ref}	reference pressure	$1 \cdot 10^5$	Pa
$c_{1,pilot}$	physical ignition delay scaling factor	0.1	—
$c_{1,main}$	physical ignition delay scaling factor	10	—
$c_{2,pilot}$	chemical ignition delay scaling factor	$4 \cdot 10^{-5}$	—
$c_{2,main}$	chemical ignition delay scaling factor	$5 \cdot 10^{-5}$	—
$c_{3,pilot}$	cylinder pressure exponential factor	-1.2	—
$c_{3,main}$	cylinder pressure exponential factor	-1.2	—
$c_{4,pilot}$	rel. A/F ratio exponential factor	0.2	—
$c_{4,main}$	rel. A/F ratio exponential factor	0.2	—
$T_{A,pilot}$	activation temperature	6000	K
$T_{A,main}$	activation temperature	6600	K

A.2.4 Premixed Combustion

- Immediately after SOC:

$$\frac{dm_{f,zone}}{d\varphi} = \min\left\{\frac{dm_{f,zone,1}}{d\varphi}, \frac{dm_{f,zone,2}}{d\varphi}\right\} \quad (A.47)$$

- At the beginning of the premixed combustion:

$$\frac{dm_{f,zone,1}}{d\varphi} = \frac{1}{1 + \lambda_{zone}(\varphi) \cdot \chi_{st} \cdot (1 + r_{zone}(\varphi))} \cdot \frac{dm_b}{d\varphi} \quad (A.48)$$

$$\frac{dm_b}{d\varphi} = \frac{1}{\omega_e} \cdot \rho_u(\varphi) \cdot s_{turb}(\varphi) \cdot A_{flame}(\varphi) \quad (A.49)$$

- Flame surface:

$$A_{flame}(\varphi) = 4 \cdot \pi \cdot \left[\int \frac{dr_{flame}}{d\varphi} \cdot d\varphi\right]^2 \cdot n_{noz} \quad (A.50)$$

$$\frac{dr_{flame}}{d\varphi} = \frac{1}{\omega_e} \cdot s_{turb}(\varphi) \cdot K_{flame}(\varphi) \quad (A.51)$$

$$K_{flame}(\varphi) = \frac{\frac{\rho_u(\varphi)}{\rho_b(\varphi)}}{\left(\frac{\rho_u(\varphi)}{\rho_b(\varphi)} - 1\right) \cdot x_b(\varphi) + 1} \quad (A.52)$$

- Combustion advancement factor:

$$x_b(\varphi) = \frac{m_f(\varphi)}{m_{inj}(\varphi)} \quad (A.53)$$

- At the end of the premixed combustion:

$$\frac{dm_{f,zone,2}}{d\varphi} = c_{pre} \cdot K_{pre}(\varphi) \cdot \frac{1}{\omega_e} \cdot \frac{1}{\tau_{pre}(\varphi)} \cdot m_{f,avail,zone}(\varphi) \quad (A.54)$$

$$K_{pre}(\varphi) = 3 \cdot \frac{1}{\Lambda_{zone}^2(\varphi)} \quad (A.55)$$

A.2. Heat Release

- Available evaporated fuel mass in the premixed zone:

$$m_{f,avail,zone}(\varphi) = m_{vap,zone}(\varphi) - m_{f,zone}(\varphi) \qquad (A.56)$$

- Characteristic premixed time:

$$\tau_{pre}(\varphi) = \frac{l_{zone}(\varphi)}{s_{turb}(\varphi)} \qquad (A.57)$$

- Characteristic length of the premixed zone:

$$l_{zone}(\varphi) = \frac{d_{zone}(\varphi)}{2} \qquad (A.58)$$

- Turbulent flame velocity:

$$s_{turb}(\varphi) = s_{lam}(\varphi) \cdot \left[1 + 1.6 \cdot \left(\frac{c_m}{s_{lam}(\varphi)}\right)^{0.8}\right] \qquad (A.59)$$

- Mean piston speed:

$$c_m = \frac{V_d \cdot \omega_e}{\pi} \qquad (A.60)$$

- Laminar flame speed (Rhodes and Keck):

$$s_{lam}(\varphi) = s_{lam,0}(\varphi) \cdot K_{lam}(\varphi) \cdot \left(\frac{T_{cyl}(\varphi)}{T_{ref}}\right)^\gamma \cdot \left(\frac{p_{cyl}(\varphi)}{p_{ref}}\right)^\delta \qquad (A.61)$$

$$s_{lam,0}(\varphi) = c_1 + c_2 \cdot \left(\frac{1}{\lambda_{zone}(\varphi)} - \frac{1}{\lambda_{max}}\right)^2 \qquad (A.62)$$

$$K_{lam}(\varphi) = 1 - 2.1 \cdot r_{zone}(\varphi) \qquad (A.63)$$

$$\gamma = 2.18 - 0.8 \cdot \left(\frac{1}{\lambda_{zone}(\varphi)} - 1\right) \qquad (A.64)$$

$$\delta = -0.16 + 0.22 \cdot \left(\frac{1}{\lambda_{zone}(\varphi)} - 1\right) \qquad (A.65)$$

Appendix A. Combustion Model: Equations and Parameters

- Accounting for all zones:

$$\frac{dm_{f,pre}}{d\varphi} = n_{noz} \cdot \frac{dm_{f,zone}}{d\varphi} \qquad (A.66)$$

$$m_{f,avail,pre}(\varphi) = n_{noz} \cdot m_{f,avail,zone}(\varphi) \qquad (A.67)$$

$$m_{vap,pre}(\varphi) = n_{noz} \cdot m_{vap,zone}(\varphi) \qquad (A.68)$$

param.	physical meaning	value	unit
V_d	cylinder displacement	$0.538 \cdot 10^{-3}$	m^3
χ_{st}	stoichiometric A/F ratio	14.67	–
n_{noz}	number of nozzle holes	6	–
c_{pre}	premixed combustion scaling factor	1.3	–
c_1	laminar velocity additive factor	0.276	–
c_2	laminar velocity multiplicative factor	-0.47	–
p_{ref}	laminar velocity reference pressure	$0.98 \cdot 10^5$	Pa
T_{ref}	laminar velocity reference temperature	298	K
λ_{max}	laminar velocity maximum rel. A/F ratio	0.91	–

A.2.5 Diffusion Combustion

- Converted fuel mass:

$$\frac{dm_{f,diff}}{d\varphi} = c_{diff} \cdot \frac{1}{\omega_e} \cdot \frac{1}{\tau_{diff}(\varphi)} \cdot m_{f,avail}(\varphi) \qquad (A.69)$$

- Characteristic diffusion time:

$$\tau_{diff}(\varphi) = \frac{l_{diff}(\varphi)}{u'(\varphi)} \qquad (A.70)$$

- Characteristic diffusion length:

$$l_{diff}(\varphi) = \sqrt[3]{\frac{V_{cyl}(\varphi)}{\lambda(\varphi) \cdot n_{noz}}} \qquad (A.71)$$

- Turbulence intensity:

$$u'(\varphi) = \sqrt{c_{back} \cdot c_m^2 + c_{kin} \cdot w_e^2 \cdot k_{tot}(\varphi)} \qquad (A.72)$$

A.2. Heat Release

- Mean piston speed:
$$c_m = \frac{V_d \cdot w_e}{\pi} \qquad (A.73)$$

- Injection energy:
$$\frac{dk_{inj}}{d\varphi} = \frac{1}{2} \cdot \frac{dm_{inj}}{d\varphi} \cdot \left(\frac{u_{dr,liq}}{w_e}\right)^2 \cdot \frac{1}{m_{cyl}(\varphi)} \qquad (A.74)$$

- Total turbulent kinetic energy:
$$\frac{dk_{tot}}{d\varphi} = -c_{diss} \cdot \frac{1}{w_e} \cdot \frac{1}{l_i} \cdot k_{tot}^{\frac{3}{2}}(\varphi) + c_{inj} \cdot \frac{dk_{inj}}{d\varphi} \qquad (A.75)$$

- Turbulent integral length:
$$l_i = d_{noz,eff} \qquad (A.76)$$

- Available fuel mass:
$$m_{f,avail}(\varphi) = m_{vap,diff}(\varphi) - m_{f,diff}(\varphi) \qquad (A.77)$$

- Evaporated fuel mass during diffusion:
$$m_{vap,diff}(\varphi) = m_{vap,tot}(\varphi) - m_{vap,pre}(\varphi) \qquad (A.78)$$

param.	physical meaning	value	unit
V_d	cylinder displacement	$0.538 \cdot 10^{-3}$	m^3
n_{noz}	number of nozzle holes	6	—
d_{noz}	nozzle hole diameter	$169 \cdot 10^{-6}$	m
c_{diff}	diffusion combustion scaling factor	1.0	—
c_{back}	background turb. intensity scaling factor	2.5	—
c_{kin}	turbulent kinetic energy scaling factor	0.25	—
c_{diss}	turbulence dissipation scaling factor	0.04	—
c_{inj}	injection-induced turb. scaling factor	0.3	—

A.2.6 Superposition of Premixed and Diffusion Combustion

- Empirical factor:

$$F_{pre/diff}(\varphi) = \left(\frac{m_{f,pre}(\varphi)}{m_{f,avail,pre}(\varphi)}\right)^{c_{pre/diff}} \quad (A.79)$$

- Final heat release rate:

$$\frac{dm_f}{d\varphi} = \frac{dm_{f,pre}}{d\varphi} + F_{pre/diff}(\varphi) \cdot \frac{dm_{f,diff}}{d\varphi} \quad (A.80)$$

param.	physical meaning	value	unit
$c_{pre/diff}$	superposition exponential delay factor	0.1	—

A.3 Single-Zone Engine Process

A.3.1 Cylinder Process

- Mass conservation:

$$\frac{dm_{cyl}}{d\varphi} = \frac{dm_f}{d\varphi} + \frac{dm_{in}}{d\varphi} + \frac{dm_{out}}{d\varphi} \quad (A.81)$$

- Energy conservation:

$$\frac{dU}{d\varphi} = \frac{dQ_f}{d\varphi} + \frac{dQ_w}{d\varphi} + \frac{dW}{d\varphi} + h_{in}\frac{dm_{in}}{d\varphi} + h_{out}\frac{dm_{out}}{d\varphi} \quad (A.82)$$

- Internal energy:

$$\frac{dU}{d\varphi} = u \cdot \frac{dm_{cyl}}{d\varphi} + m_{cyl} \cdot \frac{du}{d\varphi} \quad (A.83)$$

$$\frac{du}{d\varphi} = \frac{\partial u}{\partial T_{cyl}} \cdot \frac{dT_{cyl}}{d\varphi} + \frac{\partial u}{\partial \lambda_{cyl}} \cdot \frac{d\lambda_{cyl}}{d\varphi} \quad (A.84)$$

A.3. Single-Zone Engine Process

- Heat released:
$$\frac{dQ_f}{d\varphi} = H_f \cdot \frac{dm_f}{d\varphi} \quad (A.85)$$

- Work:
$$\frac{dW}{d\varphi} = -p_{cyl}(\varphi) \cdot \frac{dV_{cyl}}{d\varphi} \quad (A.86)$$

- Relative air/fuel ratio and its derivative:
$$\lambda_{cyl}(\varphi) = \frac{m_{cyl,air}(\varphi)}{\chi_{st} \cdot m_{cyl,f}(\varphi)} \quad (A.87)$$

$$\frac{d\lambda_{cyl}}{d\varphi} = \frac{1}{\chi_{st} \cdot m_{cyl,f}(\varphi)} \cdot \left(\frac{dm_{cyl,air}}{d\varphi} - \frac{m_{cyl,air}(\varphi)}{m_{cyl,f}(\varphi)} \cdot \frac{dm_f}{d\varphi} \right) \quad (A.88)$$

- Global relative air/fuel ratio, calculated at end of combustion:
$$\lambda_{gl} = \lambda_{cyl}(\varphi = \varphi_{EVO}) \quad (A.89)$$

- Piston motion:
$$V_{cyl}(\varphi) = V_d \cdot \left(\frac{1}{\epsilon - 1} + \zeta(\varphi) \right) \quad (A.90)$$

$$\zeta(\varphi) = \frac{1}{2} \cdot \left[(1 - \cos(\varphi - \pi)) + \left(\frac{r}{l} \cdot \sin^2(\varphi - \pi) \right) \right] \quad (A.91)$$

- Crankshaft radius:
$$r = \frac{l_{str}}{2} \quad (A.92)$$

If the effect of cylinder blow-by would also be considered, the mass flow due to blow-by would be:

$$\frac{dm_{bb}}{d\varphi} = \frac{1}{\omega_e} \cdot \mu_{bb} \cdot A_{bb} \cdot \frac{p_{cyl}(\varphi)}{\sqrt{R \cdot T_{cyl}(\varphi)}} \quad (A.93)$$

$$A_{bb} = \pi \cdot d_{bore} \cdot h_{bb} \quad (A.94)$$

144 Appendix A. Combustion Model: Equations and Parameters

param.	physical meaning	value	unit
H_f	fuel lower heating value	$42.6 \cdot 10^6$	J/kg
χ_{st}	stoichiometric A/F ratio	14.67	–
V_d	cylinder displacement	$0.538 \cdot 10^{-3}$	m^3
ϵ	compression ratio	16.4	–
d_{bore}	bore diameter	$88 \cdot 10^{-3}$	m
l_{str}	stroke length	$88.4 \cdot 10^{-3}$	m
l	con-rod length	$149 \cdot 10^{-3}$	m
μ_{bb}	blow-by discharge coefficient	$0.1 \cdot 10^{-3}$	–
h_{bb}	cylinder-piston distance	$1 \cdot 10^{-3}$	m

A.3.2 Cylinder Wall Heat Loss

- Wall heat loss:

$$\frac{dQ_w}{d\varphi} = \frac{1}{\omega_e} \cdot A_w(\varphi) \cdot q_{w,gc}(\varphi) \quad (A.95)$$

- Heat flow due to gas convection:

$$q_{w,gc}(\varphi) = \alpha(\varphi) \cdot (T_{cyl}(\varphi) - T_w(T_e)) \quad (A.96)$$

- Cylinder wall surface:

$$\begin{align}
A_w(\varphi) &= A_{bore} + A_{cyl} + A_{str}(\varphi) & (A.97)\\
A_{bore} &= \frac{\pi}{4} \cdot d_{bore}^2 & (A.98)\\
A_{cyl} &= A_{bore} - 2 \cdot (A_{v,in} + A_{v,ex}) & (A.99)\\
A_{str}(\varphi) &= d_{bore} \cdot \pi \cdot \zeta(\varphi) & (A.100)\\
\zeta(\varphi) &= \frac{1}{2} \cdot \left[(1 - \cos(\varphi - \pi)) + \left(\frac{r}{l} \cdot \sin^2(\varphi - \pi)\right)\right] & (A.101)
\end{align}$$

- Crankshaft radius:

$$r = \frac{l_{str}}{2} \quad (A.102)$$

- Woschni approach:

$$\begin{align}
\alpha(\varphi) &= 127.93 \cdot d_{bore}^{-0.2} \cdot p_{cyl}^{0.8}(\varphi) \cdot T_{cyl}^{-0.53}(\varphi) \cdot \nu^{0.8}(\varphi) & (A.103)\\
\nu(\varphi) &= C_1(\varphi) \cdot c_m + \tilde{\nu}(\varphi) & (A.104)
\end{align}$$

A.3. Single-Zone Engine Process

- Where:

$$\tilde{\nu}(\varphi) = \begin{cases} \tilde{\nu}_1 & \text{if } \tilde{\nu}_1 \leq \tilde{\nu}_2 \\ \tilde{\nu}_2 & \text{if } \tilde{\nu}_1 > \tilde{\nu}_2 \end{cases}$$

$$\tilde{\nu}_1(\varphi) = C_2 \cdot \frac{V_d \cdot T_{IVC}}{p_{IVC} \cdot V_{IVC}} \cdot (p_{cyl}(\varphi) - \tilde{p}_{cyl}(\varphi)) \quad \text{(A.105)}$$

$$\tilde{\nu}_2(\varphi) = 2 \cdot C_1(\varphi) \cdot C_3(\varphi) \cdot c_m \cdot \left(\frac{V_c}{V_{cyl}(\varphi)}\right)^2 \quad \text{(A.106)}$$

$$C_1(\varphi) = 2.28 + 0.308 \cdot \frac{c_u}{c_m} \quad \text{(A.107)}$$

$\Rightarrow$ high-pressure cycle

$$C_1(\varphi) = 6.18 + 0.417 \cdot \frac{c_u}{c_m} \quad \text{(A.108)}$$

$\Rightarrow$ gas exchange phase

$$C_2 = 3.24 \cdot 10^{-3} \, \frac{m}{s \cdot K} \quad \text{(A.109)}$$

$\Rightarrow$ diesel engines with direct injection

$$C_3(\varphi) = 1 - 1.2 \cdot e^{-0.65 \cdot \lambda_{cyl}(\varphi)} \quad \text{(A.110)}$$

- Motored cylinder pressure:

$$\tilde{p}_{cyl}(\varphi) = \left(\frac{p_{IVC} \cdot V_{IVC}}{V_{cyl}(\varphi)}\right)^{1.3} \quad \text{(A.111)}$$

- Mean piston speed:

$$c_m = \frac{V_d \cdot \omega_e}{\pi} \quad \text{(A.112)}$$

- Compression volume:

$$V_c = V_d \cdot \frac{1}{\epsilon - 1} \quad \text{(A.113)}$$

- Cylinder wall temperature:

$$T_w(T_e) = T_w^0 \cdot \left(\frac{T_e}{T_e^0}\right)^c \quad \text{(A.114)}$$

If the effect of particle radiation would also be considered, the wall heat loss equation would be:

$$\frac{dQ_w}{d\varphi} = \frac{1}{\omega_e} \cdot A_w(\varphi) \cdot (q_{w,gc}(\varphi) + q_{w,pr}(\varphi)), \quad \text{(A.115)}$$

where the heat flow due to particle radiation is defined according to [11] as:

$$q_{w,pr}(\varphi) = \sigma_{pr} \cdot \frac{\left[\left(\frac{c_T \cdot T_b(\varphi)}{100}\right)^4 - \left(\frac{T_w(T_e)}{100}\right)^4\right] \cdot \left(\frac{V_b(\varphi)}{V_{cyl}(\varphi)}\right)^{\frac{2}{3}}}{\frac{1}{\epsilon_0 - \frac{\lambda_b - \lambda_r}{c_\epsilon}} + \left(\frac{1}{\epsilon_w} - 1\right) \cdot \left(\frac{V_b(\varphi)}{V_{cyl}(\varphi)}\right)^{\frac{2}{3}}}. \quad \text{(A.116)}$$

param.	physical meaning	value	unit
V_d	cylinder displacement	$0.538 \cdot 10^{-3}$	m^3
ϵ	compression ratio	16.4	—
d_{bore}	bore diameter	$88 \cdot 10^{-3}$	m
l_{str}	stroke length	$88.4 \cdot 10^{-3}$	m
l	con-rod length	$149 \cdot 10^{-3}$	m
$A_{v,in}$	intake valve area	$71.63 \cdot 10^{-3}$	m^2
$A_{v,ex}$	exhaust valve area	$63.35 \cdot 10^{-3}$	m^2
c_u	cylinder turbulence factor	9	—
T_w^0	reference cylinder wall temperature	453	K
T_e^0	reference engine temperature	363	K
c	exponential factor	1	—
σ_{pr}	black body radiation constant	$5.67 \cdot 10^{-8}$	$W/(m^2 \cdot K^4)$
c_T	empirical factor	0.9	—
c_ϵ	empirical factor	1.0	—
ϵ_0	empirical factor	0.8	—
ϵ_w	empirical factor	0.85	—

A.3.3 Air and Exhaust Gas Properties

- Specific enthalpy:

$$h(T) = R \cdot T \cdot \left(a_0 + a_1 \cdot T + a_2 \cdot T^2 + a_3 \cdot T^3 + a_4 \cdot T^4\right) \quad (A.117)$$

- Specific internal energy:

$$u(T) = h(T) - R \cdot T \quad (A.118)$$

- Specific heat at constant pressure:

$$c_p(T) = \frac{\partial h}{\partial T} \quad (A.119)$$

- Specific heat at constant volume:

$$c_v(T) = \frac{\partial u}{\partial T} = c_p(T) - R \quad (A.120)$$

- Isentropic exponent:

$$\kappa(T) = \frac{c_p(T)}{c_v(T)} \quad (A.121)$$

- Coefficients for the specific enthalpy calculation (a fuel with 86% C-mass fraction is assumed):

component	a_0 [−]	a_1 [−]	a_2 [−]	a_3 [−]	a_4 [−]
air	3.432	$1.307 \cdot 10^{-4}$	$1.074 \cdot 10^{-7}$	$-3.429 \cdot 10^{-11}$	$2.840 \cdot 10^{-15}$
exhaust gas	3.467	$4.107 \cdot 10^{-4}$	$4.295 \cdot 10^{-8}$	$-2.714 \cdot 10^{-11}$	$2.549 \cdot 10^{-15}$

- Gas constant (a fuel with 86% C-mass fraction is assumed):

component	R [J/(kg · K)]
air	287.0
exhaust gas	286.6

Appendix A. Combustion Model: Equations and Parameters

Cylinder Charge Composition

- Air and fuel composition:

$$\frac{dm_{cyl,air}}{d\varphi} = \frac{dm_{air,gex}}{d\varphi} \tag{A.122}$$

$$\frac{dm_{cyl,f}}{d\varphi} = \frac{dm_f}{d\varphi} + \frac{dm_{f,gex}}{d\varphi} \tag{A.123}$$

- Air mass fraction:

$$\begin{aligned}
X_{air}(\varphi) &= \frac{m_{cyl,air}(\varphi) - \chi_{st} \cdot m_{cyl,f}(\varphi)}{m_{cyl}(\varphi)} = \\
&= \frac{m_{cyl,air}(\varphi) - \chi_{st} \cdot m_{cyl,f}(\varphi)}{m_{cyl,air}(\varphi) + m_{cyl,f}(\varphi)} = \\
&= \chi_{st} \cdot \frac{\lambda_{cyl}(\varphi) - 1}{1 + \lambda_{cyl}(\varphi) \cdot \chi_{st}}
\end{aligned} \tag{A.124}$$

- Specific internal energy:

$$\begin{aligned}
u(T) &= u_{air}(T) \cdot X_{air}(\varphi) + u_{eg}(T) \cdot (1 - X_{air}(\varphi)) = \\
&= \frac{u_{air}(T) \cdot \chi_{st} \cdot (\lambda_{cyl}(\varphi) - 1) + u_{eg}(T) \cdot (1 + \chi_{st})}{1 + \lambda_{cyl}(\varphi) \cdot \chi_{st}}
\end{aligned} \tag{A.125}$$

- Derivative of the specific internal energy:

$$\frac{\partial u}{\partial \lambda_{cyl}} = (u_{air}(T) - u_{eg}(T)) \cdot \frac{\chi_{st} \cdot (1 + \chi_{st})}{(1 + \lambda_{cyl}(\varphi) \cdot \chi_{st})^2} \tag{A.126}$$

param.	physical meaning	value	unit
χ_{st}	stoichiometric A/F ratio	14.67	–

A.3. Single-Zone Engine Process

A.3.4 Exhaust Gas Recirculation

- EGR rate:
$$X_{EGR} = \frac{\dot{m}_{EGR}}{\dot{m}_{inj} + \dot{m}_{air} + \dot{m}_{EGR}} \qquad (A.127)$$

- EGR mass flow:
$$\dot{m}_{EGR} = \dot{m}_{in} - \dot{m}_{air} \qquad (A.128)$$

- Theoretical mass flow entering the cylinders:
$$\dot{m}_{in,th} = \frac{p_{in} \cdot (V_d \cdot n_{cyl})}{R \cdot T_{in}} \cdot \frac{\omega_e}{4\pi} \qquad (A.129)$$

- Real mass flow entering the cylinders:
$$\dot{m}_{in} = \dot{m}_{in,th} \cdot \eta_{vol} \qquad (A.130)$$

- Volumetric efficiency:
$$\eta_{vol} = \eta_{vol,p}(p_{in}, p_{out}) \cdot \eta_{vol,\omega}(\omega_e) + \\ + \Delta\eta_{vol,T}(T_{ic}, T_{in}) \qquad (A.131)$$

- Pressure-dependent part:
$$\eta_{vol,p}(p_{in}, p_{out}) = \frac{\epsilon - \left(\frac{p_{out}}{p_{in}}\right)^{\frac{1}{\kappa}}}{\epsilon - 1} \qquad (A.132)$$

- Engine speed-dependent part:
$$\eta_{vol,\omega}(\omega_e) = c_1 + c_2 \cdot \omega_e + c_3 \cdot \omega_e^2 + c_4 \cdot \omega_e^3 \qquad (A.133)$$

- Intake temperature correction:
$$\Delta\eta_{vol,T}(T_{ic}, T_{in}) = c_5 \cdot \left(\frac{T_{in} - T_{ic}}{T_{ic}}\right)^{c_6} \qquad (A.134)$$

param.	physical meaning	value	unit
V_d	cylinder displacement	$0.538 \cdot 10^{-3}$	m^3
ϵ	compression ratio	16.4	–
n_{cyl}	number of cylinders	4	–
c_1	engine speed effect constant factor	0.6401	–
c_2	engine speed effect proportional factor	0.0018	s/rad
c_3	engine speed effect quadratic factor	$-2.9492 \cdot 10^{-6}$	$(s/rad)^2$
c_4	engine speed effect cubic factor	$-1.8533 \cdot 10^{-10}$	$(s/rad)^3$
c_5	intake temp. corr. proportional factor	1.2305	–
c_6	intake temp. corr. exponential factor	1.2196	–

A.3.5 Gas Exchange

- Flow through intake valves:

$$\frac{dm_{in}}{d\varphi} = \frac{1}{\omega_e} \cdot \mu\sigma\left(\frac{h_v}{d_v}\right) \cdot A_v \cdot \frac{\tilde{p}_{in}}{\sqrt{R \cdot \tilde{T}_{in}}} \cdot \Psi \quad (A.135)$$

$$\Psi = \sqrt{\frac{2\kappa}{\kappa - 1}\left(\Pi_v^{\frac{2}{\kappa}} - \Pi_v^{\frac{\kappa+1}{\kappa}}\right)} \quad (A.136)$$

$$\Pi_v = \min\left\{\max\left[\frac{p_{cyl}}{\tilde{p}_{in}}, \left(\frac{2}{\kappa+1}\right)^{\frac{\kappa}{\kappa-1}}\right], 1\right\} \quad (A.137)$$

- Flow through intake filling port:

$$\frac{d\tilde{m}_{in,f}}{d\varphi} = \frac{1}{\omega_e} \cdot c_d \cdot c_{IPSO} \cdot A_p \cdot \frac{p_{in}}{\sqrt{R \cdot T_{in}}} \cdot \Psi \quad (A.138)$$

$$\Psi = \sqrt{\frac{2\kappa}{\kappa - 1}\left(\Pi_v^{\frac{2}{\kappa}} - \Pi_v^{\frac{\kappa+1}{\kappa}}\right)} \quad (A.139)$$

$$\Pi_v = \min\left\{\max\left[\frac{\tilde{p}_{in}}{p_{in}}, \left(\frac{2}{\kappa+1}\right)^{\frac{\kappa}{\kappa-1}}\right], 1\right\} \quad (A.140)$$

A.3. Single-Zone Engine Process

- Flow through intake swirl port:

$$\frac{d\tilde{m}_{in,s}}{d\varphi} = \frac{1}{\omega_e} \cdot c_d \cdot A_p \cdot \frac{p_{in}}{\sqrt{R \cdot T_{in}}} \cdot \Psi \quad \text{(A.141)}$$

$$\Psi = \sqrt{\frac{2\kappa}{\kappa - 1} \left(\Pi_v^{\frac{2}{\kappa}} - \Pi_v^{\frac{\kappa+1}{\kappa}} \right)} \quad \text{(A.142)}$$

$$\Pi_v = \min\left\{ \max\left[\frac{\tilde{p}_{in}}{p_{in}}, \left(\frac{2}{\kappa + 1} \right)^{\frac{\kappa}{\kappa-1}} \right], 1 \right\} \quad \text{(A.143)}$$

- Intake ports mass receiver:

$$\frac{d\tilde{p}_{in}}{d\varphi} = \frac{\kappa \cdot R}{V_{p,in}} \cdot \left[\frac{d\tilde{m}_{in}}{d\varphi} \cdot T_{in} - \frac{dm_{in}}{d\varphi} \cdot \tilde{T}_{in} - \frac{dQ_{w,in}}{d\varphi} \cdot \frac{1}{c_p} \right] \quad \text{(A.144)}$$

- Intake ports energy receiver:

$$\frac{d\tilde{T}_{in}}{d\varphi} = \frac{\tilde{T}_{in}}{\tilde{p}_{in}} \cdot \left[\frac{d\tilde{p}_{in}}{d\varphi} - \frac{R \cdot \tilde{T}_{in}}{V_{p,in}} \cdot \left(\frac{d\tilde{m}_{in}}{d\varphi} - \frac{dm_{in}}{d\varphi} \right) \right] \quad \text{(A.145)}$$

- Intake ports wall heat losses:

$$\frac{dQ_{w,in}}{d\varphi} = \frac{1}{\omega_e} \cdot A_{p,in} \cdot q_{w,in}(\varphi) \quad \text{(A.146)}$$

$$q_{w,in} = \alpha_{p,in}(\varphi) \cdot \left(\tilde{T}_{in} - T_{w,in} \right) \quad \text{(A.147)}$$

$$\alpha_{p,in}(\varphi) = 2.152 \cdot \left(1 - 0.756 \cdot \frac{h_{v,in}}{d_{v,in}} \right) \cdot$$
$$\cdot \left(\frac{d\tilde{m}_{in}}{d\varphi} \cdot \omega_e \right)^{0.68} \cdot T_{in}^{0.28} \cdot d_{p,in}^{-1.68} \quad \text{(A.148)}$$

Appendix A. Combustion Model: Equations and Parameters

- Flow through exhaust valves:

$$\frac{dm_{out}}{d\varphi} = \frac{1}{\omega_e} \cdot \mu\sigma\left(\frac{h_v}{d_v}\right) \cdot A_v \cdot \frac{p_{cyl}}{\sqrt{R \cdot T_{cyl}}} \cdot \Psi \quad \text{(A.149)}$$

$$\Psi = \sqrt{\frac{2\kappa}{\kappa - 1}\left(\Pi_v^{\frac{2}{\kappa}} - \Pi_v^{\frac{\kappa+1}{\kappa}}\right)} \quad \text{(A.150)}$$

$$\Pi_v = \min\left\{\max\left[\frac{\tilde{p}_{out}}{p_{cyl}}, \left(\frac{2}{\kappa+1}\right)^{\frac{\kappa}{\kappa-1}}\right], 1\right\} \quad \text{(A.151)}$$

- Flow through exhaust ports:

$$\frac{d\tilde{m}_{out}}{d\varphi} = \frac{1}{\omega_e} \cdot c_d \cdot A_p \cdot \frac{\tilde{p}_{out}}{\sqrt{R \cdot \tilde{T}_{out}}} \cdot \Psi \quad \text{(A.152)}$$

$$\Psi = \sqrt{\frac{2\kappa}{\kappa - 1}\left(\Pi_v^{\frac{2}{\kappa}} - \Pi_v^{\frac{\kappa+1}{\kappa}}\right)} \quad \text{(A.153)}$$

$$\Pi_v = \min\left\{\max\left[\frac{p_{out}}{\tilde{p}_{out}}, \left(\frac{2}{\kappa+1}\right)^{\frac{\kappa}{\kappa-1}}\right], 1\right\} \quad \text{(A.154)}$$

- Exhaust ports mass receiver:

$$\frac{d\tilde{p}_{out}}{d\varphi} = \frac{\kappa \cdot R}{V_{p,ex}} \cdot \left[\frac{d\tilde{m}_{out}}{d\varphi} \cdot T_{cyl}(\varphi) - \frac{dm_{out}}{d\varphi} \cdot \tilde{T}_{out} - \frac{dQ_{w,ex}}{d\varphi} \cdot \frac{1}{c_{p,ex}}\right] \quad \text{(A.155)}$$

- Exhaust ports energy receiver:

$$\frac{d\tilde{T}_{out}}{d\varphi} = \frac{\tilde{T}_{out}}{\tilde{p}_{out}} \cdot \left[\frac{d\tilde{p}_{out}}{d\varphi} - \frac{R \cdot \tilde{T}_{out}}{V_{p,ex}} \cdot \left(\frac{dm_{out}}{d\varphi} - \frac{d\tilde{m}_{out}}{d\varphi}\right)\right] \quad \text{(A.156)}$$

A.3. Single-Zone Engine Process

- Exhaust ports wall heat losses:

$$\frac{dQ_{w,ex}}{d\varphi} = \frac{1}{\omega_e} \cdot A_{p,ex} \cdot q_{w,ex}(\varphi) \quad \text{(A.157)}$$

$$q_{w,ex}(\varphi) = \alpha_{intern}(\varphi) \cdot \left(\tilde{T}_{out} - T_{w,ex}(\varphi)\right) \quad \text{(A.158)}$$

$$q^0_{w,ex}(\varphi) = \alpha_{extern}(\varphi) \cdot (T_{w,ex}(\varphi) - T_{cool}) \quad \text{(A.159)}$$

$$\frac{dT_{w,ex}}{d\varphi} = \frac{A_{p,ex}}{m_{p,ex} \cdot c_{p,ex}} \cdot \left(q_{w,ex}(\varphi) - q^0_{w,ex}(\varphi)\right) \quad \text{(A.160)}$$

$$\alpha_{intern}(\varphi) = 1.785 \cdot \left(1 - 0.797 \cdot \frac{h_{v,ex}}{d_{v,ex}}\right) \cdot$$

$$\cdot \left(\frac{d\tilde{m}_{out}}{d\varphi} \cdot \omega_e\right)^{0.50} \cdot T^{0.41}_{cyl}(\varphi) \cdot d^{-1.50}_{p,ex} \quad \text{(A.161)}$$

$$\alpha_{extern}(\varphi) = 18.6 + 20.76 \cdot \left(\frac{T_{w,ex}(\varphi) + T_{cool}}{2}\right)^{0.50} \cdot$$

$$\cdot \left(\frac{T_{w,ex}(\varphi) - T_{cool}}{D_{p,ex}}\right)^{0.25} \quad \text{(A.162)}$$

- Variation of the air mass:

$$\frac{dm_{air,gex}}{d\varphi} = \frac{dm_{in}}{d\varphi} \cdot x_{in} + \frac{dm_{out}}{d\varphi} \cdot x_{cyl}(\varphi) \quad \text{(A.163)}$$

- Variation of the burned fuel mass:

$$\frac{dm_{f,gex}}{d\varphi} = \frac{dm_{in}}{d\varphi} \cdot (1 - x_{in}) + \frac{dm_{out}}{d\varphi} \cdot (1 - x_{cyl}(\varphi)) \quad \text{(A.164)}$$

- Air mass fraction in the intake manifold:

$$x_{in} = 1 - \frac{X_{EGR}}{1 + \lambda_{gl} \cdot \chi_{st}} \quad \text{(A.165)}$$

- Air mass fraction in the cylinder:

$$x_{cyl}(\varphi) = \frac{m_{cyl,air}(\varphi)}{m_{cyl}(\varphi)} \quad \text{(A.166)}$$

A.4 Reaction Temperature

- Temperature difference:
$$T_b(\varphi) - T_u(\varphi) = A^*(\lambda_{gl}) \cdot B(\varphi) \tag{A.167}$$

- Empirical function:
$$B(\varphi) = \frac{K - \int_{SOC}^{\varphi} (p(\varphi) - \tilde{p}(\varphi)) \cdot m_b(\varphi) \cdot d\varphi}{K} \tag{A.168}$$
$$K = \int_{SOC}^{EVO} (p(\varphi) - \tilde{p}(\varphi)) \cdot m_b(\varphi) \cdot d\varphi \tag{A.169}$$

- Maximum temperature difference between burned and unburned zone at SOC:
$$A^*(\lambda) = A \cdot \frac{1.2 + (\lambda_{gl} - 1.2)^C}{2.2 \cdot \lambda_r} \tag{A.170}$$

- Motored cylinder pressure:
$$\tilde{p}_{cyl}(\varphi) = \left(\frac{p_{IVC} \cdot V_{IVC}}{V_{cyl}(\varphi)} \right)^{1.3} \tag{A.171}$$

- Mass in the reaction zone:
$$m_b(\varphi) = m_f(\varphi) \cdot (\lambda_r \cdot \chi_{st} + 1) \tag{A.172}$$

- Pressure:
$$p_b(\varphi) = p_{cyl}(\varphi) \tag{A.173}$$
$$p_u(\varphi) = p_{cyl}(\varphi) \tag{A.174}$$

- Volume and mass conservation:
$$V_{cyl}(\varphi) = V_b(\varphi) + V_u(\varphi) \tag{A.175}$$
$$m_{cyl}(\varphi) = m_b(\varphi) + m_u(\varphi) \tag{A.176}$$

A.5. Emissions

- Gas equation (assumption: ideal gas):

$$p_b(\varphi) \cdot V_b(\varphi) = m_b(\varphi) \cdot R_b(\varphi) \cdot T_b(\varphi) \quad (A.177)$$
$$p_u(\varphi) \cdot V_u(\varphi) = m_u(\varphi) \cdot R_u(\varphi) \cdot T_u(\varphi) \quad (A.178)$$

- Density of the burned and unburned zone:

$$\rho_b(\varphi) = \frac{m_b(\varphi)}{V_b(\varphi)} \quad (A.179)$$

$$\rho_u(\varphi) = \frac{m_u(\varphi)}{V_u(\varphi)} \quad (A.180)$$

param.	physical meaning	value	unit
χ_{st}	stoichiometric A/F ratio	14.67	–
A	engine specific constant (with EGR)	1465	K
$\bar{A}$	engine specific constant (without EGR)	1520	K
C	constant	0.15	–
λ_r	rel. A/F ratio in the reaction zone	1.0	–

A.5 Emissions

A.5.1 Nitrogen Oxides

NO

- Reactions:

$$CO + \frac{1}{2}O_2 \rightleftharpoons CO_2 \;\Rightarrow\; K_{p,1} = \frac{p_{CO_2}}{p_{CO}\sqrt{p_{O_2}}} \quad (A.181)$$

$$H_2 + \frac{1}{2}O_2 \rightleftharpoons H_2O \;\Rightarrow\; K_{p,2} = \frac{p_{H_2O}}{p_{H_2}\sqrt{p_{O_2}}} \quad (A.182)$$

$$OH + \frac{1}{2}H_2 \rightleftharpoons H_2O \;\Rightarrow\; K_{p,3} = \frac{p_{H_2O}}{p_{OH}\sqrt{p_{H_2}}} \quad (A.183)$$

$$\frac{1}{2}N_2 + \frac{1}{2}O_2 \rightleftharpoons NO \;\Rightarrow\; K_{p,4} = \frac{p_{NO}}{\sqrt{p_{N_2} \cdot p_{O_2}}} \quad (A.184)$$

$$N_2 + \frac{1}{2}O_2 \rightleftharpoons N_2O \;\Rightarrow\; K_{p,5} = \frac{p_{N_2O}}{p_{N_2}\sqrt{p_{O_2}}} \quad (A.185)$$

Appendix A. Combustion Model: Equations and Parameters

- Dissociations:

$$H_2 \rightleftharpoons H + H \Rightarrow K_{p,6} = \frac{p_H}{\sqrt{p_{H_2}}} \quad (A.186)$$

$$O_2 \rightleftharpoons O + O \Rightarrow K_{p,7} = \frac{p_O}{\sqrt{p_{O_2}}} \quad (A.187)$$

$$N_2 \rightleftharpoons N + N \Rightarrow K_{p,8} = \frac{p_N}{\sqrt{p_{N_2}}} \quad (A.188)$$

- Equilibrium reaction constants (from [63]):

i	Dimension of K_p and A	A	B [−]	E_a [J/mol]
1	$[m/\sqrt{N}]$	$3.445 \cdot 10^{-10}$	0.6885	−289114
2	$[m/\sqrt{N}]$	$2.415 \cdot 10^{-5}$	−0.2421	−247233
3	$[m/\sqrt{N}]$	$1.509 \cdot 10^{-6}$	−0.1118	−288361
4	$[-]$	6.047	−0.0322	91169
5	$[m/\sqrt{N}]$	$3.769 \cdot 10^{-9}$	0.6125	78283
6	$[\sqrt{N}/m]$	$1.318 \cdot 10^{4}$	0.4056	218405
7	$[\sqrt{N}/m]$	$1.566 \cdot 10^{5}$	0.2126	249785
8	$[\sqrt{N}/m]$	$7.980 \cdot 10^{4}$	0.2920	474884

- Arrhenius approach:

$$K_{p,i} = A_i \cdot T_b^{B_i} \cdot e^{-\frac{E_{a,i}}{R \cdot T_b}} \quad (A.189)$$

A.5. Emissions

- Atomic ratios (computed according to the composition of air, i.e., 21% O_2 and 79% N_2, the composition of the diesel fuel C_mH_n, and the relative A/F ratio of the reaction zone λ_r):

$$\begin{aligned} \frac{p_O}{p_N} &= \frac{2 \cdot p_{CO_2} + 2 \cdot p_{O_2} + p_{CO} + p_{H_2O} + p_O + p_{NO} + p_{OH}}{2 \cdot p_{N_2} + p_N + p_{NO}} = \\ &= \frac{2 \cdot 0.21 \cdot \lambda_r \cdot \chi_{st}}{2 \cdot 0.79 \cdot \lambda_r \cdot \chi_{st}} = \\ &= 0.2658 \end{aligned} \qquad (A.190)$$

$$\begin{aligned} \frac{p_H}{p_C} &= \frac{2 \cdot p_{H_2O} + 2 \cdot p_{H_2} + p_H + p_{OH}}{p_{CO_2} + p_{CO}} = \\ &= \frac{n}{m} \end{aligned} \qquad (A.191)$$

$$\begin{aligned} \frac{p_O}{p_C} &= \frac{2 \cdot p_{CO_2} + 2 \cdot p_{O_2} + p_{CO} + p_{H_2O} + p_O + p_{NO} + p_{OH}}{p_{CO_2} + p_{CO}} = \\ &= \frac{2 \cdot 0.21 \cdot \lambda_r \cdot \chi_{st}}{\frac{4m}{n+4m} \cdot 0.21 \cdot \chi_{st}} = \\ &= \frac{n + 4 \cdot m}{2 \cdot m} \cdot \lambda_r \end{aligned} \qquad (A.192)$$

- Pressure balance (Dalton's law):

$$\sum_i p_i = p \qquad (A.193)$$

- Extended Zeldovich mechanism:

$$\begin{aligned} N_2 + O &\rightleftharpoons NO + N & \Rightarrow j = 1 & \qquad (A.194) \\ O_2 + N &\rightleftharpoons NO + O & \Rightarrow j = 2 & \qquad (A.195) \\ N + OH &\rightleftharpoons NO + H & \Rightarrow j = 3 & \qquad (A.196) \end{aligned}$$

Appendix A. Combustion Model: Equations and Parameters

- Forward reaction kinetic constants (from [63]):

j	A^f [m³/(mol·s)]	B^f [−]	E_a^f [J/mol]
1	$4.93 \cdot 10^7$	0.0472	$3.1627 \cdot 10^5$
2	148	1.5	$2.3765 \cdot 10^4$
3	$4.22 \cdot 10^7$	0	0

- Backward reaction kinetic constants (from [63]):

j	A^b [m³/(mol·s)]	B^b [−]	E_a^b [J/mol]
1	$1.6 \cdot 10^7$	0	0
2	12.5	1.612	$1.5769 \cdot 10^5$
3	$6.76 \cdot 10^8$	−0.212	$2.0644 \cdot 10^5$

- Arrhenius approach:

$$k_j^{f,b} = A_j^{f,b} \cdot T_b^{B_j^{f,b}} \cdot e^{-\frac{E_{a,j}^{f,b}}{R \cdot T_b}} \tag{A.197}$$

- At equilibrium (values saved in maps as a function of cylinder pressure, temperature of the reaction zone, and relative A/F ratio in the reaction zone):

$$R_{1e} = k_1^f [N_2]_e [O]_e = k_1^b [NO]_e [N]_e \tag{A.198}$$

$$R_{2e} = k_2^f [O_2]_e [O]_e = k_2^b [NO]_e [O]_e \tag{A.199}$$

$$R_{3e} = k_3^f [OH]_e [N]_e \tag{A.200}$$

$$K_e = \frac{R_{1e}}{R_{2e} + R_{3e}} \tag{A.201}$$

- Final equation:

$$\frac{d[NO]}{d\varphi} = \frac{1}{\omega_e} \cdot 2 \cdot \left(1 - \nu^2(\varphi)\right) \cdot \frac{R_{1e}(p_{cyl}, T_b, \lambda_r)}{1 + \nu(\varphi) \cdot K_e(p_{cyl}, T_b, \lambda_r)} - \xi(\varphi) \tag{A.202}$$

- Normalized concentration:

$$\nu(\varphi) = \frac{[NO](\varphi)}{[NO]_e(p_{cyl}, T_b, \lambda_r)} \tag{A.203}$$

A.5. Emissions

- Concentration variation due to volume variation:

$$\xi(\varphi) = \frac{[NO](\varphi)}{V_b(\varphi)} \cdot \frac{dV_b}{d\varphi} \qquad (A.204)$$

- Transformation from mol/m^3 to ppm:

$$NO_{ppm}(\varphi) = \frac{n_{NO}(\varphi)}{n_{eg}(\varphi)} \qquad (A.205)$$

$$n_{NO}(\varphi) = [NO](\varphi) \cdot V_b(\varphi) \qquad (A.206)$$

$$n_{eg}(\varphi) = \frac{p_{cyl}(\varphi) \cdot V_{cyl}(\varphi)}{\tilde{R} \cdot T_{cyl}(\varphi)} \qquad (A.207)$$

NO_2

- NO_2 formation:

$$NO + HO_2 \rightleftharpoons NO_2 + OH \qquad (A.208)$$
$$H + O_2 + M \rightleftharpoons HO_2 + M \qquad (A.209)$$

- NO_2 destruction:

$$NO_2 + H \rightleftharpoons NO + OH \qquad (A.210)$$
$$NO_2 + O \rightleftharpoons NO + O_2 \qquad (A.211)$$

- Empirical model for the NO_2/NO ratio:

$$\frac{[NO_2]}{[NO]} = c_1 + c_2 \cdot [NO]^{-1} \qquad (A.212)$$

param.	physical meaning	value	unit
$\tilde{R}$	universal gas constant	8.314	$J/(mol \cdot K)$
χ_{st}	stoichiometric A/F ratio	14.67	–
m	number of C atoms in the fuel	12	–
n	number of H atoms in the fuel	24	–
λ_r	rel. A/F ratio in the reaction zone	1.0	–
c_1	NO_2/NO ratio constant parameter	0.1990	–
c_2	NO_2/NO ratio proportional parameter	24.5338	ppm

A.5.2 Soot

Although this work is mainly oriented to the modeling of the NO_x formation, a model for the soot formation is implemented as well. This process is not fully understood, see for example [32]. The basic model used here has been proposed in [37] and [73]. A more detailed formulation, based on the optimization of the combustion model parameters performed by means of evolutionary algorithms, has been proposed in [27], [40], and [118].

- Soot balance:

$$\frac{dm_{soot}}{d\varphi} = \frac{dm_{soot,form}}{d\varphi} - \frac{dm_{soot,oxi}}{d\varphi} \qquad (A.213)$$

- Soot formation:

$$\frac{dm_{soot,form}}{d\varphi} = A_{form} \cdot \frac{dm_{f,diff}}{d\varphi} \cdot \left(\frac{p_{cyl}(\varphi)}{p_{ref}}\right)^{c_1} \cdot e^{-\frac{T_{A,form}}{T_{cyl}(\varphi)}} \qquad (A.214)$$

- Soot oxidation:

$$\frac{dm_{soot,oxi}}{d\varphi} = A_{oxi} \cdot \frac{1}{\omega_e} \cdot \frac{1}{\tau_{diff}(\varphi)} \cdot m_{soot}^{c_2}(\varphi) \cdot \left(\frac{p_{O_2}(\varphi)}{p_{O_2,ref}}\right)^{c_3} \cdot e^{-\frac{T_{A,oxi}}{T_{cyl}(\varphi)}} \qquad (A.215)$$

- Oxygen partial pressure:

$$p_{O_2}(\varphi) = \frac{0.21 \cdot n_{air}(\varphi) \cdot \tilde{R} \cdot T_{cyl}(\varphi)}{V_{cyl}(\varphi)} \qquad (A.216)$$

$$n_{air}(\varphi) = \frac{m_{cyl,air}(\varphi) - \chi_{st} \cdot m_{inj}(\varphi)}{M_{air}} \qquad (A.217)$$

A.5. Emissions

param.	physical meaning	value	unit
$\tilde{R}$	universal gas constant	8.314	$J/(mol \cdot K)$
M_{air}	molar weight of air	$28.84 \cdot 10^{-3}$	kg/mol
χ_{st}	stoichiometric A/F ratio	14.67	–
A_{form}	formation factor	1000	–
$T_{A,form}$	activation temperature formation	21300	K
p_{ref}	reference pressure	$1 \cdot 10^5$	Pa
A_{oxi}	oxidation factor	700	–
$T_{A,oxi}$	activation temperature oxidation	1500	K
$p_{O_2,ref}$	oxygen reference pressure	$0.21 \cdot p_{ref}$	Pa
c_1	empirical parameter	1	–
c_2	empirical parameter	1	–
c_3	empirical parameter	1.5	–

Appendix B

The Extended Kalman Filter Algorithm

B.1 Description and Assumptions

A comprehensive introduction to the extended Kalman filter as an identification tool can be found in, [13], [14] and [58].

Consider the following linear discrete time-varying system, whose formulation can be used for both the NO_x model parameter adaptation and the fault estimation:

$$\begin{cases} x(k+1) = A(k) \cdot x(k) + B(k) \cdot u(k) + B_v(k) \cdot v(k) \\ y(k) = C \cdot x(k) + q(k) \end{cases}$$

The system is corrupted by the input noise v and the measurement noise q. A convenient way to model the noise sources is to assume that they are white, independent, Gaussian-distributed, and with zero-means. The white noise assumption implies that the covariance matrices of both the input and the measurement noise vectors are diagonal, the independence assumption that the cross-covariance matrix is zero, the "gaussianity" that the estimator is statistically optimal, and the zero-means assumption that the errors in system and measurements are random. Summarizing:

$$\begin{aligned}
E\{v(k)\} &= 0 & &\Rightarrow \text{zero-mean assumption} \\
E\{q(k)\} &= 0 & &\Rightarrow \text{zero-mean assumption} \\
E\{v(k) \cdot v^T(k)\} &= R_v(t) & &\Rightarrow \text{diagonal, white noise assumption} \\
E\{q(k) \cdot q^T(k)\} &= R_q(t) & &\Rightarrow \text{diagonal, white noise assumption} \\
E\{v(k) \cdot q^T(k)\} &= 0 & &\Rightarrow \text{independence assumption}
\end{aligned}$$

However, in practical applications both noise vectors are unknown and in many cases these assumptions do not describe the reality adequately.

In common applications of the EKF technique, the state vector x contains some states that are not observable. Thus the task is to compute an estimate $\hat{x}$ of the state vector.

B.2   Algorithm

The complete EKF algorithm for a linear system is shown here. First, at each time step, the estimation residuals are computed on the basis of actual measurement data contained in the vector y:

$$r(k) = y(k) - C \cdot \hat{x}(k|k-1).$$

The "measurement update" or "correction step" is then performed in order to compute the filter-computed residual covariance matrix Q, the Kalman-gain matrix L, the state estimate vector $\hat{x}$ and the state covariance matrix Σ:

$$\begin{aligned}
Q(k) &= C \cdot \Sigma(k|k-1) \cdot C^T + R_q \\
L(k) &= \Sigma(k|k-1) \cdot C^T \cdot Q^{-1}(k) \\
\hat{x}(k|k) &= \hat{x}(k|k-1) + L(k) \cdot r(k) \\
\Sigma(k|k) &= [I - L(k) \cdot C] \cdot \Sigma(k|k-1) \cdot [I - L(k) \cdot C]^T + L(k) \cdot R_q \cdot L^T(k) \\
&= \Sigma(k|k-1) - L(k) \cdot C \cdot \Sigma(k|k-1).
\end{aligned}$$

Successively, the "time update" or "extrapolation step" is performed in order to propagate the state estimate and the state covariance matrix.

B.2. Algorithm

The matrix B_v is the input noise matrix. It is chosen to be equal to the system matrix B:

$$\hat{x}(k+1|k) = A(k) \cdot \hat{x}(k|k) + B(k) \cdot u(k)$$
$$\Sigma(k+1|k) = A(k) \cdot \Sigma(k|k) \cdot A^T(k) + B_v(k) \cdot R_v \cdot B_v^T(k).$$

The rate of convergence for the parameter estimation could be enhanced following the approach proposed in [96]. However, in this work this solution is not implemented.

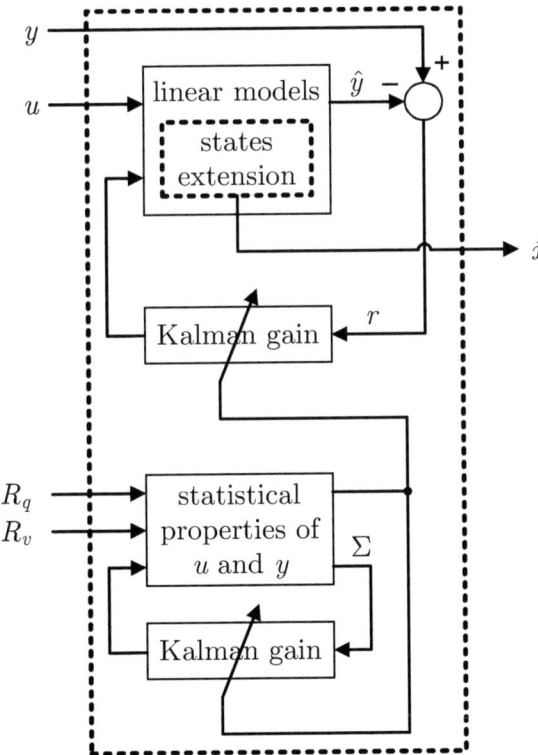

Figure B.1: Schematic representation of the EKF algorithm.

B.3 Tuning

Theoretically, the covariance matrices R_v and R_q can be obtained by measurements. In practice, their elements are set as constant, and both matrices are used as tuning parameters in order to adjust the performance of the EKF. The criterion for choosing adequate values for the elements of the two matrices is the following. If R_q is chosen large and R_v small, L becomes small, meaning that the model is trusted more than the measurements and thus that the state estimates are adapted slowly. This is the preferred adaptation strategy, since the frequency dynamics which are not reflected by the models are filtered out. On the other hand, the EKF has to deliver the state estimation as quickly as possible. Thus a compromise has to be found when designing EKF.

Since the matrix R_q describes the uncertainty of the measurement signals, in this work its choice is based on the accuracy specifications of the λ and NO_x sensors. The matrix R_v is chosen small enough in order to ensure a slow state estimation.

B.4 Observability and Excitation

An important prerequisite for the correct operation of the EKF is observability. An exhaustive study on this argument can be found in [43], [67], and [76]. For the case of this work, the observability condition specified in [67] is satisfied if the discrete observability matrix O has full rank n at the time k:

$$O(k) = \begin{bmatrix} C \\ C \cdot A(k) \\ C \cdot A(k+1) \cdot A(k) \\ \vdots \\ C \cdot A(k+n-2) \cdot A(k+n-1) \cdot \ldots \cdot A(k) \end{bmatrix}.$$

To ensure the convergence of the EKF-based identification to the "true" parameters, the system has to be sufficiently excited. The magnitude

B.4. Observability and Excitation

of this excitation is equivalent to the number of different frequencies contained in the signals of the input vector u.

Appendix C

Instrumentation

C.1 Exhaust Gas Analyzers

Cambustion fNOx400 – Fast Response CLD Nitric Oxide Measurement System

- Related publications: [24], [85], [90].

- Measure: NO and NO_x concentration (by using a thermocatalytic converter).

- Application: static measurement for combustion model calibration and verification; dynamic measurement for control-oriented model testing.

- Quenching:
  - For 1% CO_2 in sample: $\sim 0.3\%$ loss in signal.
  - For 1% H_2O in sample: $\sim 0.7\%$ loss in signal.

- Drift: $\pm 15\%/h$.

- Response time $(10 - 90\%)$: $\sim 4\ ms$.

- Principle of measurement: chemiluminescence (CLD).

Horiba Motor Exhaust Gas Analyzer MEXA-1300FRI

- Measure: CO, CO_2 and HC concentration.
- Application: static measurement for EGR calculation.
- Measuring range:
  - CO: $0 - 12\%$ V.
  - CO_2: $0 - 16\%$ V.
  - HC: 2% V.
- Drift: $\pm 1\%/h$.
- Response time $(0 - 90\%)$: ~ 30 ms.
- Principle of measurement: nondispersive infrared analysis (NDIR).

Pierburg Golem II

- Measure: CO, CO_2, HC and NO_x concentration (by using a thermocatalytic converter).
- Application: static measurement.
- Quenching (only for NO_x):
  - For 1% CO_2 in sample: $\sim 0.1\%$ loss in signal.
  - For 1% H_2O in sample: $\sim 0.3\%$ loss in signal.
- Response time: ~ 1.5 s for CO and CO_2, ~ 2 s for HC and NO_x.
- Principle of measurement:
  - For CO and CO_2: nondispersive infrared analysis (NDIR).
  - For HC: flame ionization detection (FID).
  - For NO_x: chemiluminescence (CLD).

C.2. Sensors

AVL 483 Micro Soot Sensor

- Measure: soot concentration.
- Application: static and dynamic measurement.
- Response time: $\sim 1.0\ s$.
- Principle of measurement: photo-acustic soot sensor (PASS).

C.2 Sensors

Bosch Wide-Range λ Sensor LSU 4.9

- Related publications: [95].
- Measure: λ.
- Application: on-line.
- Type: ZrO_2-based multilayer sensor element.
- Measuring range: $\lambda = 0.65$ to air.
- Accuracy new:
  - At $\lambda = 0.8$: $\pm 1.3\%$.
  - At $\lambda = 1$: $\pm 0.6\%$.
  - At $\lambda = 1.7$: $\pm 2.9\%$.
- Accuracy after 3000 h of operation:
  - At $\lambda = 0.8$: $\pm 5\%$.
  - At $\lambda = 1$: $\pm 0.8\%$.
  - At $\lambda = 1.7$: $\pm 8.8\%$.
- Response time: $\sim 70\ ms$.

- Principle of measurement: oxygen concentration proportional to oxygen ion current.

Siemens-NGK Smart NO_x Sensor

- Related publications: [99].
- Measure: NO_x concentration.
- Application: on-line.
- Type: ZrO_2-based multilayer sensor element.
- Measuring range: $0 - 1500\ ppm$.
- Accuracy: $\pm 10\%$.
- Response time $(33 - 66\%)$: $\sim 750\ ms$.
- Principle of measurement: oxygen concentration proportional to oxygen ion current.

Micro-Epsilon Turbocharger Speed Sensor

- Measure: turbocharger rotational speed.
- Measuring range: $0 - 400000\ rpm$.
- Principle of measurement: impedance variation due to eddy current losses.

Sensortechnics BTE6000 Pressure Transmitters

- Measure: pressure.
- Measuring range: $0-2\ bar$ (BTE6002) and $0-5\ bar$ (BTE6005).
- Response time $(10 - 90\%)$: $1\ ms$.

Bibliography

Books

[1] AA.VV.: *Automotive Handbook, 4th Edition*, Robert Bosch GmbH, 1996

[2] AA.VV.: *Dieselmotor-Management, 2. Auflage*, Robert Bosch GmbH, 1998

[3] Gertler J.J.: *Fault Detection and Diagnosis in Engineering Systems*, Marcel Dekker Inc., 1998

[4] Guzzella L., Onder C.H.: *Introduction to Modeling and Control of Internal Combustion Engine Systems*, Springer Verlag, 2004

[5] Guzzella L., Sciarretta A.: *Vehicle Propulsion Systems*, Springer Verlag, 2005

[6] Heywood J.B.: *Internal Combustion Engine Fundamentals*, McGraw-Hill, 1988

[7] Höfling T.: *Methoden zur Fehlererkennung mit Parameterschätzung und Paritätsgleichungen*, Fortschritt-Bericht VDI, Reihe 8 Mess-, Steuerungs- und Regelungstechnik, Nr. 546, VDI Verlag, Düsseldorf, 1999

[8] Isermann R.: *Identifikation dynamischer Systeme − Band I + II*, Springer Verlag, 1988

[9] Isermann R.: *Überwachung und Fehlerdiagnose*, VDI Verlag, 1994

[10] Isermann R.: *Fault-Diagnosis Systems − An Introduction from Fault Detection to Fault Tolerance*, Springer Verlag, 2006

[11] Kleinschmidt W.: *Instationäre Wärmeübertragung in Verbrennungsmotoren*, Fortschritt-Bericht VDI, Reihe 12 Verkehrstechnik/Fahrzeugtechnik, Nr. 383, VDI Verlag, Düsseldorf, 1999

[12] Kolar J.: *Stickstoffoxide und Luftreinhaltung*, Springer Verlag, 1990

[13] Ljung L., Söderström T.: *Theory and Practice of Recursive Identification*, The MIT Press, 1983

[14] Ljung L.: *System Identification: Theory for the User*, Prentice-Hall, 1987

[15] Merker G.P., Schwarz C.: *Technische Verbrennung*, Teubner Verlag, Wiesbaden, 2001

[16] Merker G.P., Schwarz C., Stiesch G., Otto F.: *Verbrennungsmotoren − Simulation der Verbrennung und Schadstoffbildung*, Teubner Verlag, Wiesbaden, 2004

[17] Patton R.J., Frank P.M., Clark R.N.: *Issues of Fault Diagnosis for Dynamic Systems*, Springer Verlag, 2000

[18] Pischinger R., Klell M., Sams T.: *Thermodynamik der Verbrennungskraftmaschine*, Springer Verlag, 2002

[19] Stiesch G.: *Modeling Engine Spray and Combustion Processes*, Springer Verlag, 2003

[20] Turns S.R.: *An Introduction to Combustion − Concepts and Applications*, McGraw-Hill, 1996

Ph.D. Theses

[21] Allmendinger K.: *Modellbildung und modellbasierte Estimation thermodynamischer Prozessgrössen am Beispiel eines Dieselmotors*, Dissertation Universität-Gesamthochschule Siegen, 2002

[22] Ammann M.: *Modellbasierte Regelung des Ladedrucks und der Abgasrückführung beim aufgeladenen PKW-Common-Rail-Dieselmotor*, Diss. ETH Nr. 15166, ETH Zürich, 2003

[23] Amstutz A.: *Geregelte Abgasrückführung zur Senkung der Stickoxid- und Partikelemissionen beim Dieselmotor mit Comprex-Aufladung*, Diss. ETH Nr. 9421, ETH Zürich, 1991

[24] Baltisberger S.: *Entwicklung eines NO-Messgeräts zur arbeitsspielauflösenden Abgasanalyse an Verbrennungsmotoren*, Diss. ETH Nr. 11593, ETH Zürich, 1996

[25] Barba C.: *Erarbeitung von Verbrennungskennwerten aus Indizierdaten zur verbesserten Prognose und rechnerischen Simulation des Verbrennungsablaufes bei Pkw-DE-Dieselmotoren mit Common-Rail-Einspritzung*, Diss. ETH Nr. 14276, ETH Zürich, 2001

[26] Bargende M.: *Ein Gleichungsansatz zur Berechnung der instationären Wandwärmeverluste im Hochdruckteil von Ottomotoren*, Dissertation TU Darmstadt, 1991

[27] Bertola A.G.: *Technologies for Lowest NOx and Particulate Emissions in DI-Diesel Engine Combustion − Influence of Injection Parameters, EGR and Fuel Composition*, Diss. ETH No. 15373, ETH Zürich, 2003

[28] Brand D.: *Control-Oriented Modeling of NO Emissions of SI Engines*, Diss. ETH No. 16037, ETH Zürich, 2005

[29] Ganser M.: *Akkumuliereinspritzung: theoretische und experimentelle Untersuchung eines elektronisch gesteuerten Dieseleinspritzsystems für Personenwagenmotoren*, Diss. ETH Nr. 7462, ETH Zürich, 1984

[30] Heider G., *Rechenmodell zur Vorausrechnung der NO-Emission von Dieselmotoren*, Dissertation TU München, 1996

[31] Kozuch P.: *Ein phänomenologisches Modell zur kombinierten Stickoxid- und Russberechnung bei direkteinspritzenden Dieselmotoren*, Dissertation Universität Stuttgart, 2004

[32] Kunte S.: *Untersuchungen zum Einfluss von Brennstoffstruktur und -sauerstoffgehalt auf die Russbildung und -oxidation in laminaren Diffusionsflammen*, Diss. ETH Nr. 15003, ETH Zürich, 2003

[33] Lootsma T.F.: *Observer-based Fault Detection and Isolation for Nonlinear Systems*, Ph.D. Thesis, Aalborg University, 2001

[34] Nyberg M.: *Model Based Fault Diagnosis – Methods, Theory and Automotive Engine Applications*, Dissertation No. 591, Linköping University, 1999

[35] Schär C.M.: *Control of a Selective Catalytic Reduction Process*, Diss. ETH No. 15221, ETH Zürich, 2003

[36] Schernewski R.: *Modellbasierte Regelung ausgewählter Antriebssystemkomponenten im Kraftfahrzeug*, Dissertation Universität Karlsruhe, 1999

[37] Schubiger R.A.: *Untersuchungen zur Russbildung und -oxidation in der dieselmotorischen Verbrennung: Thermodynamische Kenngrössen, Verbrennungsanalyse und Mehrfarbenendoskopie*, Diss. ETH Nr. 14445, ETH Zürich, 2001

Bibliography

[38] Torkzadeh D.D.: *Echtzeitsimulation der Verbrennung und modellbasierte Reglersynthese am Common-Rail-Dieselmotor*, Dissertation Universität Karlsruhe, 2003

[39] Vogt R.: *Beitrag zur rechnerischen Erfassung der Stickoxidbildung im Dieselmotor*, Dissertation Universität Stuttgart, 1975

[40] Warth M.: *Comparative Investigation of Mathematical Methods for Modeling and Optimization of Common-Rail DI Diesel Engines*, Diss. ETH No. 16357, ETH Zürich, 2005

Articles

[41] Amstutz A., Del Re L.R.: *EGO Sensor Based Robust Output Control of EGR in Diesel Engines*, IEEE Transactions on Control Systems Technology, Vol. 3, No. 1, 1995, pp. 39-48

[42] Brand D., Onder C.H., Guzzella L.: *Virtual NO Sensor for Spark-ignition Engines*, International Journal of Engine Research, Vol. 8, No. 2, 2007, pp. 221-240

[43] Fitts J.M.: *On the Observability of Nonlinear Systems with Applications to Nonlinear Regression Analysis*, Information Sciences, Vol. 4, 1972, pp. 129-156

[44] Gertler J.J., Costin M., Fang X., Hira R., Kowalczuk Z., Luo Q.: *Model-Based On-Board Fault Detection and Diagnosis for Automotive Engines*, Control Engineering Practice, Vol. 1, No. 1, 1993, pp. 3-17

[45] Gertler J.J., Costin M., Fang X., Kowalczuk Z., Kunwer M., Monajemy R.: *Model Based Diagnosis for Automotive Engines – Algorithm Development and Testing on a Production Vehicle*, IEEE Transactions on Control Systems Technology, Vol. 3, No. 1, 1995, pp. 61-69

[46] Guzzella L., Amstutz A.: *Control of Diesel Engines*, IEEE Control Systems Magazine, Vol. 18, No. 5, 1998, pp. 53-71

[47] Hafner M., Schüler M., Nelles O., Isermann R.: *Fast Neural Networks for Diesel Engine Control Design*, Control Engineering Practice, Vol. 8, 2000, pp. 1211-1221

[48] Heider G., Woschni G., Zeilinger K.: *2-Zonen Rechenmodell zur Vorausrechnung der NO-Emission von Dieselmotoren*, MTZ 11, 1998, pp. 770-775

[49] Herrmann O.E., Krüger M., Pischinger S.: *Regelung von Ladedruck und AGR-Rate als Mittel zur Emissionsregelung bei Nutzfahrzeugmotoren*, MTZ 10, 2005, pp. 806-811

[50] Isermann R.: *Model-Based Fault Detection and Diagnosis – Status and Applications*, Annual Reviews in Control, Vol. 29, 2005, pp. 71-85

[51] Jung M., Glover K., Christen U.: *Comparison of Uncertainty Parameterisations for H_∞ Robust Control of Turbocharged Diesel Engines*, Control Engineering Practice, Vol. 13, No. 1, 2005, pp. 15-25

[52] Kim Y.W., Rizzoni G., Utkin V.: *Automotive Engine Diagnosis and Control via Nonlinear Estimation*, IEEE Control Systems Magazine, Vol. 18, No. 5, 1998, pp. 84-99

[53] Kimmich F., Schwarte A., Isermann R.: *Fault Detection for Modern Diesel Engines Using Signal- and Process Model-Based Methods*, Control Engineering Practice, Vol. 13, No. 5, 2005, pp. 189-203

[54] Klingmann R., Brüggemann H.: *Der neue Vierzylinder-Dieselmotor OM611 mit Common-Rail-Einspritzung – Teil 1: Motorkonstruktion und mechanischer Aufbau*, MTZ 11, 1997, pp. 652-659

[55] Kozuch P., Grill M., Bargende M.: *Ein neuer Ansatz zur kombinierten Stickoxyd- und Russberechnung bei DI-Dieselmotoren*, Dieselmotorentechnik 2004, Vol. 656, Expert Verlag 2004, pp. 93-113

[56] Lamping M., Körfer T., Pischinger S.: *Zusammenhang zwischen Schadstoffreduktion und Verbrauch bei Pkw-Dieselmotoren mit Direkteinspritzung*, MTZ 01, 2007, pp. 50-57

[57] Leonhardt S., Müller N., Isermann R.: *Methods for Engine Supervision and Control Based on Cylinder Pressure Information*, IEEE/ASME Transactions on Mechatronics, Vol. 4, No. 3, 1999

[58] Ljung L.: *Asymptotic Behavior of the Extended Kalman Filter as a Parameter Estimator for Linear Systems*, IEEE Transactions on Automatic Control, Vol. 24, No. 1, 1979, pp. 36-50

[59] Nyberg M.: *Automatic Design of Diagnosis Systems with Application to an Automotive Engine*, Control Engineering Practice, Vol. 7, No. 8, 1999, pp. 993-1005

[60] Nyberg M.: *Model-based Diagnosis of an Automotive Engine Using Several Types of Fault Models*, IEEE Transactions on Control Systems Technology, Vol. 10, No. 5, 2002, pp. 679-689

[61] Nyberg M.: *Using Hypothesis Testing Theory to Evaluate Principles for Leakage Diagnosis of Automotive Engines*, Control Engineering Practice, Vol. 11, 2003, pp. 1263-1272

[62] Nyberg M.: *Model-based Diagnosis of the Air Path of an Automotive Diesel Engine*, Control Engineering Practice, Vol. 12, 2004, pp. 513-525

[63] Pattas K., Häfner G.: *Stickoxydbildung bei der ottomotorischen Verbrennung*, MTZ 12, 1973, pp. 397-404

[64] Payri F., Lujan J.M., Guardiola C., Rizzoni G.: *Injection Diagnosis Through Common-rail Pressure Measurement*, Proceedings

of the IMechE, Part D: Journal of Automobile Engineering, Vol. 220, No. 3, 2006, pp. 347-357

[65] Peters A., Putz W.: *Der neue Vierzylinder-Dieselmotor OM611 mit Common-Rail-Einspritzung – Teil 2: Verbrennung und Motormanagement*, MTZ 12, 1997, pp. 760-767

[66] Ramamoorthy R., Dutta P.K., Akbar S.A.: *Oxygen Sensors: Materials, Methods, Designs and Applications*, Journal of Material Science, Vol. 38, 2003, pp. 4271-4282

[67] Reif K., Günther S., Yaz E., Unbehauen R.: *Stochastic Stability of the Discrete-Time Extended Kalman Filter*, IEEE Transaction on Automatic Control, Vol. 44, No. 4, 1999, pp. 714-728

[68] Riegel J., Neumann H., Wiedenmann H.M.: *Exhaust Gas Sensors for Automotive Emission Control*, Solid State Ionics, Vol. 152-153, 2002, pp. 783-800

[69] Rizzoni G, Min P.S.: *Detection of Sensor Failures in Automotive Engines*, IEEE Transactions on Vehicular Technology, Vol. 40, No. 2, 1991

[70] Rückert J., Kinoo B., Krüger M., Schlosser A., Rake H., Pischinger S.: *Simultane Regelung von Ladedruck und AGR-Rate beim Pkw-Dieselmotor*, MTZ 11, 2001, pp. 956-965

[71] Schilling A., Alfieri E., Amstutz A., Guzzella L.: *Emissionsgeregelte Dieselmotoren*, MTZ 11, 2007, pp. 982-989

[72] Schilling A., Amstutz A., Guzzella L.: *Model-Based Detection and Isolation of Faults due to Ageing in the Air and Fuel Paths of Common-Rail DI Diesel Engines Equipped with a Lambda and a Nitrogen Oxides Sensor*, Proceedings of the IMechE, Part D: Journal of Automobile Engineering, Vol. 222, No. 1, 2008, pp. 101-118

Bibliography

[73] Schubiger R.A., Boulouchos K., Eberle M.K.: *Russbildung und Oxidation bei der dieselmotorischen Verbrennung*, MTZ 5, 2002, pp. 342-353

[74] Schwarte A., Kimmich F., Isermann R.: *Modellbasierte Fehlererkennung und -diagnose für Dieselmotoren*, MTZ 7-8, 2002, pp. 612-620

[75] Soliman A., Rizzoni G., Kim Y.W.: *Diagnosis of an Automotive Emission Control System Using Fuzzy Inference*, Control Engineering Practice, Vol. 7, 1999, pp. 209-216

[76] Sontag E.D.: *On the Observability of Polynomial Systems I: Finite-time Problems*, SIAM Journal of Control and Optimization, Vol. 17, No. 1, 1979, pp. 139-151

[77] Sun J., Kim, Y.W., Wang L.: *Aftertreatment Control and Adaptation for Automotive Lean Burn Engines with HEGO Sensors*, International Journal of Adaptive Control and Signal Processing, Vol. 18, No. 2, 2004, pp. 145-166

[78] Visser J.H., Soltis R.E.: *Automotive Exhaust Gas Sensing Systems*, IEEE Transactions on Instrumentation and Measurement, Vol 50, No. 6, 2001, pp. 1543-1550

[79] Vogt M., Müller N., Isermann R.: *On-Line Adaptation of Grid-Based Look-up Tables Using a Fast Linear Regression Technique*, ASME Journal of Dynamic Systems, Measurement, and Control, Vol. 126, 2004, pp. 732-739

[80] Warth M., Boulouchos K., Obrecht P.: *Kennfeldtaugliche Vorausberechnungen beim Dieselmotor*, MTZ 11, 2004, pp. 924-931

[81] Zeilinger K., Zitzler G.: *Vorausberechnung der Brennverläufe von Gasmotoren*, MTZ 12, 2003, pp. 1080-1089

Conference Papers

[82] Alfieri E., Amstutz A., Onder C.H., Guzzella L.: *Automatic Design and Parametrization of a Model-Based Controller Applied to the AF-Ratio Control of a Diesel Engine*, Proceedings of the IFAC Symposium on Advances in Automotive Control, 2007, Monterey Coast

[83] Ammann M., Fekete N.P., Guzzella L., Glattfelder A.H.: *Model-Based Control of the VGT and EGR in a Turbocharged Common-Rail Diesel Engine: Theory and Passenger Car Implementation*, SAE 2003-01-0357, 2003

[84] Arnold J.F., Langlois N., Chafouk H., Tremouliere G.: *Fuzzy Controller for the Air System of a Diesel Engine*, ECOSM – Rencontres Scientifiques de l'IFP, 2006, Proceedings pp. 87-93

[85] Baltisberger S., Ruhm K.: *Fast NO Measuring Device for Internal Combustion Engines*, SAE SP-1014, 1994, pp. 53-59

[86] Barba C., Burkhardt C., Boulouchos K., Bargende M.: *A Phenomenological Combustion Model for Heat Release Rate Prediction in High-Speed DI Diesel Engines with Common-Rail Injection*, SAE 2000-01-2933, 2000

[87] Bianchi G.M., Falfari S., Parotto M., Osbat G.: *Advanced Modeling of Common Rail Injector Dynamics and Comparison with Experiments*, SAE 2003-01-0006, 2003

[88] Brahma I., Rutland C.J. Foster D.E., He Y.: *A New Approach to System Level Soot Modeling*, SAE 2005-01-1122, 2005

[89] Chen S.K., Yanakiev O.: *Transient NOx Emission Reduction Using Exhaust Oxygen Concentration Based Control for a Diesel Engine*, SAE 2005-01-0372, 2005

[90] Collier T., Gregory D., Rushton M., Hands T.: *Investigation into the Performance of an Ultrafast Response NO Analyser Equipped*

with a NO2 to NO Converter for Gasoline and Diesel Exhaust NOx Measurements, SAE 2000-01-2954, 2000

[91] Del Re L.R., Langthaler P., Furtmüller C., Winkler S., Affenzeller M.: *NOx Virtual Sensor Based on Structure Identification and Global Optimization*, SAE 2005-01-0050, 2005

[92] Gangopadhay A., Matekunas F., Battison P., Szymkowicz P., Pinson J., Landsmann G.: *Control of Diesel HCCI Modes Using Cylinder Pressure-based Controls*, F2006P285, FISITA 2006

[93] Garcia-Ortiz J.V., Langthaler P., Del Re L.R.: *GPC Control of the Airpath of High Speed Diesel Engines*, Proceedings of the IEEE CCA 2006, München, pp. 2772-2777

[94] Gauthier C., Sename O., Dugard L., Meissonnier G.: *An LFT Approach to H_∞ Control Design for Diesel Engine Common Rail Injection System*, ECOSM – Rencontres Scientifiques de l'IFP, 2006, Proceedings pp. 213-219

[95] Hackel V., Schnabel C., Tiefenbach A.: *Wide Band Oxygen Sensor Electronic Control Unit (LambdaTronic)*, SAE 2005-01-0061, 2005

[96] Hill B.K., Walker B.K.: *Approximate Effect of Parameter Pseudonoise Intensity on Rate of Convergence for EKF Parameter Estimators*, Proceedings of the 30th IEEE Conference on Decision and Control, 1991, pp. 1690-1697

[97] Hoffmann K., Hessler F., Abel D.: *Rapid Control Prototyping with Dymola and Matlab for a Model Predictive Control for the Air Path of a Boosted Diesel Engine*, ECOSM – Rencontres Scientifiques de l'IFP, 2006, Proceedings pp. 25-33

[98] James S., Joshi A.A., King G.B., Meckl P.H.: *Diagnosis of Clogged Charge Air Cooler Faults in a Diesel Engine Using Singular Spectrum Analysis*, Proceedings of the 5th IFAC Sympo-

sium on Advances in Automotive Control, 2007, Monterey Coast, pp. 311-318

[99] Kato N., Nakagaki K., Ina N.: *Thick Film ZrO2 NOx Sensor*, SAE 960334, 1996

[100] Kämmer A., Liebl J., Krug C., Munk F., Reuss H. C.: *Real-Time Engine Models*, SAE 2003-01-1050, 2003

[101] Klett S., Piesche M., Heinzelmann S., Weyl H., Wiedenmann H.M., Schneider U., Diehl L., Neumann H.: *Numerical and Experimental Analysis of the Momentum and Heat Transfer in Exhaust Gas Sensors*, SAE 2005-01-0037, 2005

[102] Krug C., Liebl J., Munk F., Kämmer A., Reuss H.C.: *Physical Modelling and Use of Modern System Identification for Real-Time Simulation of Spark Ignition Engines in all Phases of Engine Development*, SAE 2004-01-0421, 2004

[103] Liu Y., Tao F., Foster D.E., Reitz R.D: *Application of a Multiple-Step Phenomenological Soot Model to HSDI Diesel Multiple Injection Modeling*, SAE 2005-01-0924, 2005

[104] Manchur T.B., Checkel M.D.: *Time Resolution Effects on Accuracy of Real-Time NOx Emissions Measurements*, SAE 2005-01-0674, 2005

[105] Maybeck P.S.: *Multiple Model Adaptive Algorithms for Detecting and Compensating Sensor and Actuator/Surface Failures in Aircraft Flight Control System*, International Journal of Robust and Nonlinear Control 9, 1999, pp. 1051-1070

[106] Merz B., Walter A., Brumm S., Kiencke U.: *Real-Time Calculation of the Nitric Oxide Formation as an Add-on for a Zero-Dimensional Model of the Diesel Combustion*, Proceedings of the 5th IFAC Symposium on Advances in Automotive Control, 2007, Monterey Coast, pp. 405-411

Bibliography 185

[107] Morselli R., Corti E., Rizzoni G.: *Energy Based Model of a Common Rail Injector*, Proceedings of the IEEE CCA 2002, Glasgow, pp. 1195-1200

[108] Nitsche R., Hanschke J., Schwarzmann D.: *Nonlinear Internal Model Control of Diesel Air Systems*, ECOSM – Rencontres Scientifiques de l'IFP, 2006, Proceedings pp. 121-131

[109] Nyberg M.: *Model Based Diagnosis of Both Sensor-Faults and Leakage in the Air-Intake System of an SI-Engine*, SAE 1999-01-0860, 1999

[110] Ortner P, Langthaler P., Garcia-Ortiz J.V., Del Re L.R.: *MPC for a Diesel Engine Air Path using an Explicit Approach for Constraint Systems*, Proceedings of the IEEE International Conference on Control Applications, 2006, München, pp. 2760-2765

[111] Rupp D., Ducard G., Shafai E., Geering H.P.: *Extended Multiple Model Adaptive Estimation for the Detection of Sensor and Actuator Faults*, Proceedings of the IEEE ECC-CDC 2005, pp. 3079-3084

[112] Schär C.M., Onder C.H., Geering H.P., Elsener M.: *Control of the Urea SCR Catalytic Converter System for a Mobile Heavy Duty Diesel Engine*, SAE 2003 Transactions, Journal of Engines, Detroit, 2004

[113] Schilling A., Amstutz A., Onder C.H., Guzzella L.: *A Real-Time Model for the Prediction of the NOx Emissions in DI Diesel Engines*, Proceedings of the IEEE International Conference on Control Applications, 2006, München, pp. 2042-2047

[114] Stotsky A.: *Data-Driven Algorithms for Engine Friction Estimation*, Proceedings of the IEEE International Conference on Control Applications, 2006, München, pp. 521-526

[115] Tao F., Liu Y., Rempelewert B.H., Foster D.E., Reitz R.D., Choi D., Miles P.C.: *Modeling the Effects of EGR and Injection Pressure on Soot Formation in a High-Speed Direct-Injection (HSDI) Diesel Engine Using a Multi-Step Phenomenological Soot Model*, SAE 2005-01-0121, 2005

[116] Traver M.L., Atkinson R.J., Atkinson C.M.: *Neural Network-Based Diesel Engine Emissions Prediction Using In-Cylinder Combustion Pressure*, SAE 1999-01-1532, 1999

[117] Upadhyay D., Van Nieuwstadt M.: *NOx Prediction in Diesel Engines for Aftertreatment Control*, IMECE 2003-41196, ASME International Mechanical Engineering Congress, 2003

[118] Warth M., Obrecht P., Bertola A.G., Boulouchos K.: *Predictive Phenomenological C.I. Combustion Modeling Optimization on the Basis of Bio-Inspired Algorithms*, SAE 2005-01-1119, 2005

[119] Woermann R.J., Theuerkauf H.J., Heinrich A.: *A Real-Time Model of a Common Rail Diesel Engine*, SAE 1999-01-0862, 1999

FVV Projects

[120] Hild O., Schlosser A., Fieweger K, Deutsch G.: *Abgasturbolader – Pkw-Dieselmotor mit Abgasturboaufladung, variabler Turbinengeometrie/Abgasrückführung*, FVV Vorhaben Nr. 651, Abschlussbericht, 1998

[121] Kimmich F., Schwarte A., Isermann R.: *Diagnosemethoden Dieselmotor – Modellgestützte präventive Diagnosemethoden (Fehlerfrüherkennung) für Dieselmotoren*, FVV Vorhaben Nr. 709, Abschlussbericht, 1999

[122] Kinoo B., Krüger M., Pischinger S., Rückert J. Schlosser A., Enning M., Rake H.: *Modellgestützte Dieselmotorregelung*

— *Modellgestützte Mehrgrössenregelung eines Pkw-Dieselmotors mit VTG-Lader und Abgasrückführung*, FVV Vorhaben Nr. 732, Abschlussbericht, 2001

Die VDM Verlagsservicegesellschaft sucht für wissenschaftliche Verlage abgeschlossene und herausragende

Dissertationen, Habilitationen, Diplomarbeiten, Master Theses, Magisterarbeiten usw.

für die kostenlose Publikation als Fachbuch.

Sie verfügen über eine Arbeit, die hohen inhaltlichen und formalen Ansprüchen genügt, und haben Interesse an einer honorarvergüteten Publikation?

Dann senden Sie bitte erste Informationen über sich und Ihre Arbeit per Email an *info@vdm-vsg.de*.

Sie erhalten kurzfristig unser Feedback!

VDM Verlagsservicegesellschaft mbH
Dudweiler Landstr. 99   Telefon +49 681 3720 174
D - 66123 Saarbrücken   Fax   +49 681 3720 1749

www.vdm-vsg.de

Die VDM Verlagsservicegesellschaft mbH vertritt

Printed by Books on Demand GmbH, Norderstedt / Germany